COMMON MINERALS AND ROCKS

COMMON MINERALS AND ROCKS

WILLIAM OTIS CROSBY

ISBN: 978-93-5614-405-7

Published: -

LECTOR HOUSE LLP
E-MAIL: lectorpublishing@gmail.com

BOSTON SOCIETY OF NATURAL HISTORY

*   *   *   *   *

GUIDES FOR SCIENCE-TEACHING
NO. XII

COMMON MINERALS AND ROCKS

BY

WILLIAM O. CROSBY

CONTENTS

INTRODUCTION.

Minerals and rocks, or the inorganic portions of the earth, constitute the proper field or subject-matter of the science of Geology. Now the inorganic earth, like an animal or plant, may be and is studied in three quite distinct ways, giving rise to three great divisions of geology, which, as will be seen, correspond closely to the main divisions of Biology.

First, we may study the forces now operating upon and in the earth—the geological agencies—such as the ocean and atmosphere, rivers, rain and frosts, earthquakes, volcanoes, hot springs, etc., and observe the various effects which they produce. We are concerned here with the dynamics of the earth; and this is the great division of *dynamical geology*, corresponding to physiology among the biological sciences.

Or, second, instead of geological causes, we may study more particularly geological effects, observing the different kinds of rocks and of rock-structure produced by the geological agencies, not only at the present time, but also during past ages. This method of study gives us the important division of *structural geology*, corresponding to anatomy and morphology.

All phenomena present two distinct and opposite aspects or phases which we call *cause* and *effect*; and so in dynamical and structural geology we are really studying the opposite sides of essentially the same classes of phenomena. In the first division we study the causes now in operation and observe their effects; and then, guided by the light of the experience thus gained, we turn to the effects produced in the past and seek to refer them to their causes.

These two divisions together constitute what is properly known as physiography; and they are both subordinate to the third great division of geology,—*historical geology*,—which corresponds to embryology.

The great object of the geologist is, by studying the geological formations in regular order, from the oldest up to the newest, to work out, in their proper sequence, the events which constitute the earth's history; and dynamical and structural geology are merely introductory chapters, the alphabet, as it were, which must be learned before we are prepared to read understandingly the grand story of the geological record.

Our work in this short course will be limited to the first two divisions,—*i.e.*, to dynamical and structural geology. We will attempt, first, a general sketch of the forces now concerned in the formation of rocks and rock-structures; and after that we will study the composition and other characteristics of the common minerals and rocks.

The scope of this work, and its relations to the whole field of geology, are more clearly indicated by the following classification of the geological sciences:—

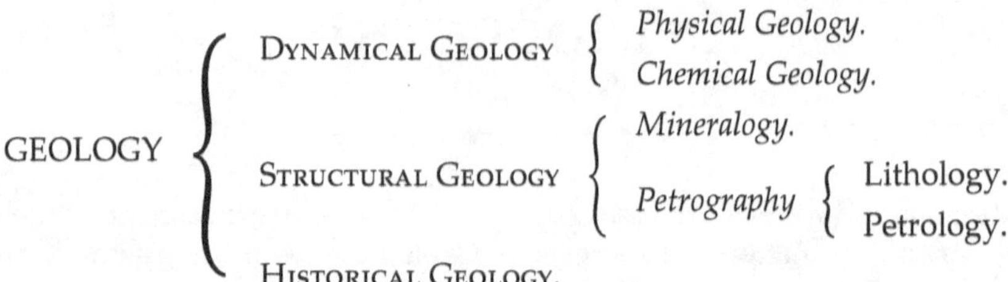

Many teachers will desire to fill in some of the details of the outline sketch presented in this Guide, and for this purpose the following works are especially recommended:—

> ELEMENTS OF GEOLOGY. By Prof. Joseph Le Conte. 1882. D. Appleton & Co., New York. Nearly 600 pages.

> MANUAL OF GEOLOGY. By Prof. J. D. Dana. Third edition. 1880. 800 pages.

> TEXT-BOOK OF GEOLOGY. By Prof. A. Geikie. 1882. Macmillan & Co., London. Nearly 1000 pages.

As a reference-book for mineralogy, the following treatise is unsurpassed:—

> TEXT-BOOK OF MINERALOGY. By Edward S. Dana. 1883. John Wiley & Sons, New York.

And, as an introduction to the study of minerals, and, through these, to the study of rocks,—

> FIRST LESSONS IN MINERALS. Science Guide No. XIII. By Mrs. E. H. Richards.

cannot be too highly recommended. Teachers will find this little primer of 46 pages invaluable with young children, and with all who have had no previous training in chemistry.

As an admirable continuation of the work begun in these pages, teachers are referred to Professor Shaler's "First Book in Geology." In this our brief sketch of the geological agencies is amplified and beautifully illustrated; and rarely have the wonderful stories of the river, ocean-beach, glacier, and volcano been told so effectively. In the chapter on the history of life on the globe the main outlines of historical geology are skillfully brought within the comprehension of beginners. The directions to teachers are fully in accord with the modern methods and ideas, and are a very valuable feature of the book.

DYNAMICAL GEOLOGY.

When we think of the ocean with its waves, tides, and currents, of the winds, and of the rain and snow, and the vast net-work of rivers to which they give rise, we realize that the energy or force manifested upon the earth's surface resides chiefly in the *air* and *water*—in the earth's fluid envelope and not in its solid crust. And it would be an easy matter to show that, with the exception of the tidal waves and currents, which of course are due chiefly to the attraction of the moon, nearly all this energy is merely the transformed heat of the sun. Now the air and water are two great geological agencies, and therefore the geological effects which they produce are traceable back to the sun.

Organic matter is another important geological agent; but all are familiar with the generalization that connects the energy exhibited by every form of life with the sun; and, besides, it is scarcely necessary to allude to the obvious fact that all animals and plants, so far at least as any display of energy is concerned, are merely differentiated portions of the earth's fluid envelope. And so, if space permitted, it might be shown that, with the exception of the tides, nearly every form of force manifested upon the earth's surface has its origin in the sun.

Of this trio of geological agencies operating upon the earth's surface and vitalized by the sun—*water, air,* and *organic matter*—the water is by far the most important, and so it is common to call these collectively the aqueous agencies. Hence we have *solar agencies* and *aqueous agencies* as synonymous terms.

The aqueous agencies include, on one side, *air* and *water,* or *inorganic* agencies; and, on the other, *animals* and *plants,* or *organic* agencies.

Let us notice briefly the operation of these, beginning with the air and water.

I. AQUEOUS AGENCIES.

1. *Air and Water, or Inorganic Agencies.*

CHEMICAL EROSION.—Attention is invited first to the specimens numbered 1, 2, 3, and 4. No. 1 is a sound, fresh piece of the rather common rock, diabase; and those who are acquainted with minerals will recognize that the light-colored grains in the rock are feldspar, and the dark, augite. This specimen came from a depth in the quarry, and has not been exposed to the action of the weather.

The second specimen differs from the first, apparently, as much as possible;

and yet, except in being somewhat finer grained, it was originally of precisely similar composition and appearance. In fact, it is a portion of the same rock, but a *weathered* portion. In this we can no longer recognize the feldspar and augite as such, but both these minerals are very much changed, while in the place of a strong, hard rock we have an incoherent friable mass, which is, externally at least, easily crushed to powder; and with the next step in the weathering, as we may readily observe in the natural ledges, the rock is completely disintegrated, forming a loose earth or soil.

We have two examples of such natural powders in the specimens numbered 3 and 4; and by washing these (especially the finer one, No. 4) with water, we can prove that they consist of an impalpable substance which we may call clay, and angular grains which we may call sand. The sand-grains are really portions of the feldspar not yet entirely changed to clay.

Thus we learn that the result of the exposure of this hard rock to the weather is that it is reduced to the condition of sand and clay. What we mean especially by the weather are *moisture* and certain constituents of the air, particularly *carbon dioxide*.

The action of the weather on the rocks is almost entirely chemical. With a very few exceptions, the principal minerals of which rocks are composed, such as feldspar, hornblende, augite, and mica, are silicates, *i.e.*, consist of silicic acid or silica combined with various bases, especially aluminum, magnesium, iron, calcium, potassium, and sodium.

Now the silica does not hold all these bases with equal strength; but carbon dioxide, in the presence of moisture, is able to take the sodium, potassium, calcium, and magnesium away from the silica in the form of carbonates, which, being soluble, are carried away by the rain-water.

The silicate of aluminum, with more or less iron, takes on water at the same time, and remains behind as a soft, impalpable powder, which is common clay.

In the case of our diabase, continued exposure to the weather would reduce the whole mass to clay. But other rocks contain grains of quartz, a hard mineral which cannot be decomposed, and it always forms sand. Certain classes of rocks, too, such as the limestones and some iron-ores, are completely dissolved by water holding carbon dioxide in solution, and nothing is left to form soil, except usually a small proportion of insoluble impurities like sand or clay.

Let us see next how these agents of decay get at the rocks. Neither water nor air can penetrate the solid rock or mineral to any considerable extent, so that practically the action is limited to surfaces, and whatever multiplies surfaces must favor decomposition.

First, we have the upper surface of the rock where it is bare, but more especially where it is covered with soil, for there it is always wet.

All rocks are naturally divided by joints into blocks, which are frequently more or less regular, and often of quite small size. Water and air penetrate into these cracks and decompose the surfaces of the blocks, and thus the field of their operations is enormously extended. These rock-blocks sometimes show very beautifully the progress of the decomposing agents from the outside inward by concentric layers or shells of rotten material, which, in the larger blocks, often envelop a nucleus of the unaltered rock.

It is interesting to observe, too, that these concentric lines of decay cut off the angles of the original blocks, so that the undecomposed nucleus, when it is found, is approximately spherical instead of cuboidal. Both these points are well illustrated by specimen No. 2; for although now nearly spherical, it was originally perfectly angular, and has become rounded by the peeling off, in concentric layers, of the decomposed material, and in most cases several of these layers are distinctly visible, like the coats of an onion. But by stripping these off we should discover, in all the larger balls at least, a solid, spheroidal nucleus, while in the smaller balls the decomposition has penetrated to the centre.

In the rocks also we find many imperfect joints and minute cracks. In cold countries these are extended and widened by the expansive power of freezing water, and thus the surfaces of decomposition become constantly greater.

Nearly all rocks suffer this chemical decomposition when exposed to the weather, but in some the decay goes on much faster than in others. Diabase is one of the rocks which decay most readily; while granite is, among common rocks, one of those that resist decay most effectually.

The caverns which are so large and numerous in most limestone countries are a splendid example of the solvent action of meteoric waters, being formed entirely by the dissolving out of the limestone by the water circulating through the joint cracks. The process must go on with extreme slowness at first, when the joints are narrow, and more rapidly as they are widened and more water is admitted. We get some idea, too, of the magnitude of the results accomplished by these silent and unobtrusive agencies when we reflect that almost all the loose earth and soil covering the solid rocks are simply the insoluble residue which carbon dioxide and water cannot remove. In low latitudes, where a warm climate accelerates the decay of the rocks, the soil is usually from 50 to 300 feet deep.

MECHANICAL EROSION. — *On the edge of the land.* —Let us trace next the *mechanical* action of water and air upon the land. First we will consider the *edge* of the land, where it is washed by the waves of the sea. Whoever has been on the shore must have noticed that the sand along the water's edge is kept in constant motion by the ebb and flow of the surf.

Where the beach is composed of gravel or shingle the motion is evident to the *ear* as well as the eye; and when the surf is strong, the rattling and grinding of the pebbles as they are rolled up and down the beach develops into a roar.

The constant shifting of the grains of sand, pebbles, and stones is, of course, attended by innumerable collisions, which are the cause of the noise. Now it is practically impossible, as we may easily prove by experiment, to knock or rub two pieces of stone together, at least so as to produce much noise, without abrading their surfaces; small particles are detached, and sand and dust are formed.

That this abrasion actually occurs in the case of the moving sand is most beautifully shown by the sandblast. We are to conclude, then, that every time a pebble, large or small, is rolled up or down the beach it becomes smaller, and some sand and dust or clay are formed which are carried off by the water.

But what are the pebbles originally? This question is not difficult. A little observation on the beach shows us that the pebbles are not all equally round and smooth, but many are more or less angular. And we soon see that it is possible to select a series showing all gradations between the most perfectly rounded forms and angular fragments of rock that are only slightly abraded on the corners. The three principal members of such a series are shown in specimens 5, 6, and 7 from the beach on Marblehead Neck; but equally instructive specimens can be obtained at many other points on our coast. It is also observable that the well-rounded pebbles are much smaller on the average than the angular blocks.

From these facts we draw the legitimate inference that the pebbles were all originally angular, and that the same abrasion which diminishes their size makes them round and smooth.

A little reflection, too, shows that the rounding of the angular fragments is a natural and necessary result of their mutual collisions; for the angles are at the same time their weakest and most exposed points, and must wear off faster than the flat or concave surfaces.

Having traced each pebble back to a larger angular rock-fragment, the question arises, Whence come these angular blocks?

Behind our gravel-beach, or at its end, we have usually a cliff of rocks. As we approach this it is distinctly observable that the angular pebbles are more numerous, larger, and more angular; and a little observation shows that these are simply the blocks produced by jointing, and that the cliff is entirely composed of them. In other words, our cliff is a mass of natural masonry, which chemical agencies, the frost, and the sea are gradually disintegrating and removing. As soon as the blocks are brought within reach of the surf their mutual collisions make them rounder and smaller; and small round pebbles, sand, and clay are the final result.

For a more complete account of the formation of pebbles, teachers are referred to the first or introductory number of this series of guides, by Prof. Hyatt, "About Pebbles."

Where the waves can drive the shingle directly against the base of the cliff, this is gradually ground away in the same manner as the loose stones themselves, sometimes forming a cavern of considerable depth, but always leaving a smooth, hard surface, which is very characteristic, and contrasts strongly with the upper portion of the cliff, which is acted on only by the rain and frost. A good example of such a pebble-carved cliff may be seen behind the beach on the sea-ward side of Marblehead Neck.

The sea acts within very narrow limits vertically, a few feet or a few yards at most; but the coast-lines of the globe (including inland lakes and seas) have an aggregate length of more than 150,000 miles. Hence it is easy to see that the amount of solid rock ground to powder in the mill of the ocean-beach annually must be very considerable.

MECHANICAL EROSION.—*On the surface of the land.*—I next ask attention to the *mechanical* action of water upon the *surface* of the land.

It is a familiar fact that after heavy rains the roadside rills carry along much sand and clay (which we know have been produced by the previous action of chemical forces), and also frequently small pebbles or gravel. It is easy to show that in all important respects the rill differs in size only from brooks and rivers; and the former afford us fine models of the systems of valleys worn out during the lapse of ages by rivers. The turbidity of rivers is often very evident, and in shallow streams we can sometimes see the pebbles rolled along by the current.

Now here, just as on the beach, the collisions of rock-fragments are attended by mutual abrasion, sand and clay are formed, and the fragments become smaller and rounder. Our series of pebbles from the beach might be matched perfectly among the river-gravel. In mountain streams especially we may often observe that pebbles of a particular kind of rock become more numerous, larger, and more angular as we proceed up stream, until we reach the solid ledge from which they were derived, showing the same gradation as the beach pebbles when followed back to the parent cliff.

The pebbles, however, not only grind each other, but also the solid rocks which form the bed of the streams in many places, and these are gradually worn away. When the rocky bed is uneven and the water is swift, pebbles collect in hollows where eddies are formed, by which they are kept whirling and turning, and the hollow is deepened to a pot-hole, while the pebbles, the river's tools, are worn out at the same time.

By these observations we learn not only that running water carries away sand and clay already formed, but that it also has great power of grinding down hard rocks to sand and clay. Of course the pulverized rock always moves in the same direction as the stream which carries it; and, in a certain sense, all streams run in one direction, viz., toward the sea. Therefore the constant tendency of the rain falling upon the land is to break up the rocks by chemical and mechanical action and transport the débris to the sea.

Rivers, as we all know, are continually uniting to form larger and larger streams; and thus the drainage of a wide area sometimes, as in the case of the Mississippi Valley, reaches the sea through a single mouth. By careful measurements made at the mouth of the Mississippi it has been shown that the 20,000,000,000,000 cubic feet of water discharged into the Gulf of Mexico annually carries with it no less than 7,500,000,000 cubic feet of sand, clay, and dissolved mineral matter; and this, spread over the whole Mississippi basin, would form a layer a little more than $\frac{1}{5000}$ of a foot in thickness. So that we may conclude that the surface of the continent is being cut down on the average about *one foot* in *five thousand* years.

We can only allude in passing to the very important geological action of water in the solid state, as in glaciers and icebergs. The moisture precipitated from the atmosphere, and falling as rain, makes ordinary rivers; but falling in the form of snow in cold regions, where more snow falls than is melted, the excess accumulates and is gradually compacted to ice, which, like water, yields to the enormous pressure of its own mass and flows toward lower levels. When the ice-river reaches the sea it breaks off in huge blocks, which float away as icebergs. Moving ice, like moving water, is a powerful agent of erosion; and the glacial marks or scratches observable upon the ledges everywhere in the Northern States and Canada attest the magnitude of the ice-action at a comparatively recent period.

We have already noticed incidentally the powerful disintegrating action of water where it freezes in the joints and pores of the rocks; and it is probable that it thus facilitates the destruction of the rocks in cold countries nearly as much as the higher temperature and greater rain-fall do in warm countries.

Our observations up to this point show us that *erosion*, by which we mean the breaking up by chemical and mechanical action of the rocks of the land and the transportation of the débris into the sea, is one great result accomplished by the inorganic aqueous agencies.

MECHANICAL DEPOSITION.—Next let us notice what becomes of all this vast amount of clay, sand, and gravel after it is washed into the ocean. By taking up a glass of turbid water from our roadside rill, and observing that as soon as the water is undisturbed the sand and clay begin to settle, we learn that the solid matter is held in suspension by the motion of the water. But it does not remain in suspension long after being washed into the sea, for otherwise the sea would, in the course of time, become turbid for long distances from shore; and it is a well-known fact that the sea-water is usually clear and free from sensible turbidity close along shore and even near the mouths of large rivers, while at a distance of only 50 or 100 miles we find the transparency of the central ocean.

Putting these facts together, we see that the ocean, nothwithstanding the ceaseless and often violent undulations of its surface, must be as a whole a vast body of still water; and to the reflecting mind the almost perfect tranquillity of the ocean is one of its most impressive features. For it is in striking contrast, in this respect, with the more mobile aerial ocean above it.

We have got hold, now, of two facts of great geological importance: (1) The débris washed off the land by waves and rivers into the still water of the ocean very soon settles to the bottom; and (2) it nearly all settles on that part of the ocean-floor near the land.

And now we have in view the second great office of the inorganic aqueous agencies,—deposition, the counterpart or complement of erosion.

The land is the great theatre of erosion and the sea of deposition; the rocks which are constantly wasting away on the former are as constantly renewed in the latter.

We will now observe the process of deposition a little more closely. Each of these two bottles contains the same amount of fine yellow clay, but in one the water is fresh, and in the other it is salt. At the beginning of the lesson, as you may have observed, I brought the clay in both bottles into suspension by violent agitation, and since then they have remained undisturbed. The main point is that the salt water has become quite clear, while the fresh water is still distinctly turbid, showing that the salt favors the rapid deposition of the clay. At the second lecture, a week later, these two bottles, yet undisturbed, were exhibited, and the fresh water seen to be still sensibly turbid. The fact is, the clay is not held in suspension wholly by the *motion* of the water; but, just as in the case of dust in the atmosphere, a small portion of the medium is condensed around or adheres to each solid particle, *i.e.,* each clay particle in our experiment has an atmosphere of water which moves with it and buoys it up. Now the effect of the salt is to diminish the adhesion of the water to the particles, *i.e.,* to diminish their atmospheres, and consequently their buoyancy. The diminished adhesion of the salt water is well shown by the smaller drops which it forms on a glass rod.

The geological importance of this principle is very great; for it is undoubtedly largely to the saltness of the sea that we owe its transparency, and the fact that the fine, clayey sediment from the land, like the coarse, is deposited near the shore.

This bottle of fresh water contains some fine gravel, coarse sand, fine sand, and clay. By agitating the water, all this material is brought into suspension. Now, suddenly placing the bottle in a state of rest, we observe that the gravel falls to the bottom almost instantly, followed quickly by the coarse sand, and very soon afterward by the fine sand; and then there appears to be a pause, the fine particles of clay all remain in suspension; but finally, when the water is quite motionless, they begin to settle; they fall very slowly, however, and the water will not be clear for hours.

This is a very instructive experiment. We learn from it:

First, that the power of the water to hold particles in suspension is inversely proportional to the size of the particles;

Second, that all materials deposited in water are assorted according to size;

Third, and this is one of the most important facts in geology, all water-deposited sediments are arranged in horizontal layers, *i.e.,* are stratified. And we have now traced to its conclusion, though very briefly, the process of the formation of one great division of *stratified* rocks,—the *mechanically-formed* or *fragmental* rocks. These are so called because the clay, sand, and gravel are, in every instance, fragments of pre-existing rocks; and because the formation, transportation, and especially the *deposition* of these fragments, are the work chiefly or entirely of mechanical forces.

CHEMICAL DEPOSITION.—It is a well-known fact that the sea holds in solution vast amounts of common salt as well as many other substances; and analyses of river-waters show that dissolved minerals derived from the chemical decomposition of the rocks of the land are being constantly carried into the sea.

Portions of the sea which are cut off from the main body, and which are gradually drying up, like the Great Salt Lake, Dead Sea, and Caspian Sea, become saturated solutions of the various dissolved minerals, and these are slowly deposited. This process is very nicely illustrated along our shores in summer, where, during storms, salt-water spray is thrown above the reach of the tides, and, collecting in hollows in the rocks, gradually dries up, leaving behind a crust of salt.

When the sea lays down matter which it held in *suspension*, we call the process *mechanical* deposition, and the result is *mechanically*-formed rocks.

But when it lays down matter which it held in *solution*, we call the process *chemical* deposition, and the result is *chemically*-formed rocks.

The principal substances which the sea deposits chemically are common salt, forming beds of rock-salt; sulphate of calcium, forming beds of gypsum; carbonate of calcium, forming beds of limestone; and the double carbonate of calcium and magnesium, forming beds of dolomite.

Inorganic deposition, like inorganic erosion, is both chemical and mechanical.

2. *Animals and Plants, or Organic Agencies.*

We turn now to the consideration of the *organic* agencies. And I will merely allude in passing to the vast importance of the fossil organic remains found in the stratified rocks as marks by which to determine the relative ages of the formations.

As regards the *destruction* of rocks—*erosion*—plants and animals are almost powerless; but in the role of *rock-makers* they play a very important part, being very efficient agents of *deposition*.

FORMATION OF COALS AND BITUMENS.—Specimen No. 8 is an example of peat from the vicinity of Boston; but just as good specimens may be obtained in thousands of places in this and other States.

The general physical conditions under which peat is formed are familiar facts. We require simply low, level land, covered with a thin sheet of water and abundant vegetation; in other words, a marsh or swamp. If plants decay on the dry land, the decomposition is complete; they are burned up by the oxygen of the air to *carbon dioxide* and *water* just as surely as if they had been thrown into a furnace, though less rapidly, and nothing is returned to the soil but what had been taken from it by the plants during their growth. But if the plants decay under water, as in a peat-marsh or bog, the decay is incomplete, and most of the carbon of the wood is left behind. Now, if this incomplete combustion of vegetable tissues takes place in a charcoal-pit, where the wood is out of contact with air from being covered with earth, we call the carbonaceous product charcoal; but if under the water of a marsh, in Nature's laboratory, we call the product peat. Peat is simply a natural charcoal; and, just as in ordinary charcoal, its vegetable origin is always perfectly evident. But when the deposit becomes thicker, and especially when it is buried under thick formations of other rocks, like sand and clay, the great pressure consolidates the peat; it becomes gradually more mineralized and shining, shows the vegetable tissues less distinctly, becomes more nearly pure carbon, and we call it in succession lignite, bituminous coal, and anthracite.

This is, briefly, the way in which all varieties of coal, as well as the more solid kinds of bitumen, like asphaltum, are formed. But the lighter forms of bitumen, such as petroleum and naphtha, are derived mainly, if not entirely, from the partial decomposition of animal tissues. These, it is well known, decay much more readily than vegetable tissues; and the water of an ordinary marsh or lake contains sufficient oxygen for their complete and rapid decomposition. In the deeper parts of the ocean, however, the conditions are very different, for recent researches have shown, contrary to the old idea, that the deep sea holds an abundant fauna. All grades of animal life, from the highest to the lowest, have need of a constant supply of oxygen. On the land vegetation is constantly returning to the air the oxygen consumed by animals, but in the abysses of the ocean vegetable life is scarce or wanting; and hence it must result that over these greater than continental areas countless myriads of animals are living habitually on short rations of oxygen, and in water well charged with carbon dioxide, the product of animal respiration. As a consequence, when these animals die their tissues do not find the oxygen essential for their perfect decomposition, and in the course of time become buried, in a half-decayed state, in the ever-increasing sediments of the ocean-floor.

It is important to observe that an abundance of organic matter decaying under water is not the only condition essential to the formation of beds of coal and bitumen; for this condition is realized in the luxuriant growth of sea-weeds fringing the coast in every quarter of the globe; and yet coals and bitumens are rarely of sea-shore origin. These organic products, even under the most favorable circumstances, accumulate with extreme slowness; far more slowly, as a rule, than the ordinary mechanical sediments, like sand and clay, with which they are mixed, and in which they are often completely lost. Consequently, although the deposition of the carbonized remains of plants and animals is taking place in nearly all seas, lakes, and marshes, it is only in those places where there is little or no mechanical sediment that they can predominate so as to build up beds pure enough to be called coal or bitumen. In all other cases we get merely more or less carbonaceous sand or clay. Now these especially favorable localities will manifestly not be often found along the seashore, where we have strewn the sand and clay brought down by rivers or washed off the land directly by the ever-active surf; but they must exist in the central portions of the ocean, where there is almost no mechanical sediment and yet an abundance of life, and in swamps and marshes, where there is scarcely sufficient water to cover the vegetation, and no waves or currents to wash down the soil from the surrounding hills.

FORMATION OF IRON-ORES. — The iron-ores are another class of rocks which are formed only through the agency of organic matter. Iron is an abundant and widespread element in the earth's crust, and, but for the intervention of life, we might say that, while there is iron everywhere, there is not much of it in any one place, since it is originally very thinly diffused. All rocks and soils contain iron, but it is mainly in the form of the peroxide, in which state it is entirely insoluble, and hence cannot be soaked out of the soil by the rain-water and concentrated by the evaporation of the water at lower levels in ponds and marshes, as a soluble substance like salt would be. If carried off with the sand and clay, by the mechanical action of water, it remains uniformly mixed with them, and there is no tendency to its separation and concentration so as to form a true iron-ore.

But what water cannot do alone is accomplished very readily when the water is aided by decaying organic matter, which is always hungry for oxygen, being, in the language of the chemist, a powerful reducing agent. The soil, in most places, has a superficial stratum of vegetable mould or half-decayed vegetation. The rainwater percolates through this and dissolves more or less of the organic matter, which is thus carried down into the sand and clay beneath and brought in contact with the ferric oxide, from which it takes a certain proportion of oxygen, reducing the ferric to the ferrous oxide. At the same time the vegetation is burned up by the oxygen thus obtained, forming carbon dioxide, which immediately combines with the ferrous oxide, forming carbonate of iron, which, being soluble under these conditions, is carried along by the water as it gradually finds its way by subterranean drainage to the bottom of the valley and emerges in a swamp or marsh.

Here one of two things will happen: If the marsh contains little or no decaying vegetation, then as soon as the ferrous carbonate brought down from the hills is exposed to the air it is decomposed, the carbon dioxide escapes, and the iron, taking on oxygen from the air, returns to its original ferric condition; and being then quite insoluble, it is deposited as a loose, porous, earthy mass, commonly known as bog-iron-ore, which becomes gradually more solid and finally even crystalline through the subsequent action of heat and pressure. When first deposited, the ferric oxide is combined with water or hydrated, and is then known as limonite (specimen No. 12); at a later period the water is expelled, and we call the ore hematite (specimen No. 13); and at a still later age it loses part of its oxygen, becomes magnetic and more crystalline, and is then known as magnetite (specimen No. 14). Thus it is seen that the iron-ores, as we pass from bog-limonite to magnetite, form a natural series similar to and parallel with that afforded by the coals as we pass from peat to graphite.

If the drainage from the hills is into a marsh containing an abundance of decaying vegetation, *i.e.*, if peat is forming there, the ferrous carbonate, in the presence of the more greedy organic matter, will be unable to obtain oxygen from the air; and as the evaporation of the water goes on, it will sooner or later become saturated with this salt, and the latter will be deposited. Here we find an explanation of a fact often observed by geologists, viz., that the carbonate iron-ores are usually associated with beds of coal.

The formation of the iron-ores, like that of the coals and bitumens, is a slow process; and the ores, like the coals, etc., will be pure only where there is a complete absence of mechanical sediment, a condition that is realized most nearly in marshes.

FORMATION OF LIMESTONE, DIATOMACEOUS EARTH, ETC.—Marine animals take from the sea-water certain mineral substances, especially silica and carbonate of calcium, to form their skeletons. Silica is used only by the lowest organisms, such as Radiolaria, Sponges, and the minute unicellular plants, Diatoms. The principal animals secreting carbonate of calcium are Corals and Mollusks. These hard parts of the organisms remain undissolved after death; and over portions of the ocean-floor where there is but little of other kinds of sediment they form the main part of the deposits, and in the course of ages build up very extensive formations which we call diatomaceous earth or tripolite, if the organisms are siliceous, or limestone if they are calcareous. A very satisfactory account of the formation of limestone on a stupendous scale by the polyps in coral reefs and islands is contained in No. IV. of this series of guides.

The rocks here considered may be, and, as we have already seen, sometimes are, deposited in a purely chemical way, without the aid of life; and it is important to observe that in no case do the organisms make the silica and carbonate of calcium of their skeletons, but they simply appropriate and reduce to the solid state what exists ready made in solution in the sea-water. These minerals, and others, as we know, are produced by the decomposition of the rocks of the land, and are being constantly carried into the sea by rivers; and, if there were no animals in the sea, these processes would still go on until the sea-water became saturated with these substances, when their precipitation as limestone, etc., would necessarily follow. Hence it is clear that all the animals do is to effect the precipitation of certain minerals somewhat sooner than it would otherwise occur; so that from a geological standpoint the differences between chemical and organic deposition are not great.

This section of our subject may be summarized as follows: Animals and plants contribute to the formation of rocks in three distinct ways:—

1. During their growth they deoxidize carbon dioxide and water, and reduce to the solid state in their tissues carbon and the permanent gases oxygen, hydrogen, and nitrogen; and after death, through the accumulation of the half-decayed tissues in favorable localities,—marshes, etc.,—these elements are added to the solid crust of the earth in the form of coal and bitumen.

2. During the decomposition, *i.e.*, oxidation, of the organic tissues, the iron existing everywhere in the soil is partially deoxidized, and, being thus rendered soluble, is removed by rain-water and concentrated in low places, forming beds of iron-ore.

3. Through the agency of marine organisms, certain mineral substances are being constantly removed from the sea-water and deposited upon the ocean floor, forming various calcareous and siliceous rocks.

I now bring our study of the aqueous or superficial agencies to a conclusion by noting once more that the great geological results accomplished by *air, water*, and *organic matter* or *life* are: (1) *Erosion*, or the wearing away of the surface of the land; and (2) *Deposition*, or the formation from the débris of the eroded land of two great classes of stratified rocks,—the mechanically formed or fragmental rocks, and the chemically and organically formed rocks.

II. IGNEOUS AGENCIES.

We pass next to a very brief consideration of operations that originate below the earth's surface. The records of deep mines and artesian wells show that the temperature of the ground always increases downwards from the surface; and the much higher temperatures of hot springs and volcanoes show that the heat continues to increase to a great depth, and is not a merely superficial phenomenon. The observed rate of increase is not uniform, but it seldom varies far from the average, which is about 1° Fahr. per 53 feet of vertical descent, or, in round numbers, 100° per mile. This rate, if continued, would give a very high temperature at points only a few miles below the surface; and until within a few years the idea was generally accepted by geologists that the increase of temperature is sensibly uniform for an indefinite distance downward; that in the central regions of the earth the temperature is far higher than anything we can conceive, and that everywhere below a depth of 20 to 40 miles the temperature is above the fusing-point of all rocks; and hence that the earth is an incandescent liquid globe covered by a thin shell or crust of cold, solid rock.

Our limited space will not permit us to enter into a discussion of the condition of the earth's interior, and I will merely point out in a few sentences the position occupied by geologists at the present time. The reasoning of Thompson has shown that the temperature cannot increase downward at a uniform rate, but at a constantly and rapidly diminishing rate; and that everywhere below a depth of 300 miles the temperature is probably sensibly the same, and nowhere, probably, above 8000° to 10,000° Fahr.

Unlike water, all rocks contract on solidifying and expand on melting, and consequently the high pressures to which they are subjected in the earth's interior—10,000,000 to 20,000,000 pounds per square inch—must raise their fusing-points enormously, and the probabilities are that they are solid, in spite of the high temperature. But Thompson and Darwin have shown us farther that the phenomena of the oceanic tides could not be what they are known to be if the earth were any less rigid than a globe of solid steel; while Hopkins has proved that the astronomical phenomena of precession and nutation could not be what they are if the earth's crust were less than 800 or 1000 miles thick. Putting these considerations together, geologists are almost universally agreed that, while the earth has an incandescent interior, it is still continuously solid from centre to circumference, with the exception of a thin plastic stratum at a depth not exceeding 40 or 50 miles, which forms the seat of volcanic action.

The earth is not only a very hot body, but it is rotating through almost absolutely cold space, and therefore must be a cooling body. But, except at the very beginning of the cooling, the loss of heat has gone on almost entirely from the interior; and since cooling means contraction, the heated interior must be constantly tending to shrink away from the cold external crust.

Of course no actual separation between the crust and interior or nucleus can take place, but there is no doubt that the crust is left unsupported to a certain extent, and it must then behave like an arch with a radius of 4000 miles, and the result is an enormous horizontal or tangential pressure.

This lateral pressure in the earth's crust is one of the most important and most generally accepted facts in geology, and lies at the bottom of many geological theories. According to what seems to me to be the most probable theory of the origin of continents and ocean-basins, they are broad upward and downward bendings or arches into which the crust is thrown by the tangential pressure. Finally, the strain becomes great enough to crush the crust along those lines where it is weakest. When the crust is thus mashed up by horizontal pressure, a mountain range is formed, the crust becomes enormously thicker, and a weak place becomes a strong one.

During the formation of mountains the stratified rocks, which were originally horizontal, are thrown into folds or arches, and tipped up at all possible angles; they are fractured and faults produced; and by the immense pressure the structure known as slaty cleavage is developed. In fact, a vast amount and variety of structures are produced during the growth of a mountain range.

These great earth-movements are not always perfectly smooth and steady, but they are accompanied by slipping or crushing now and then; and, as a result of the shock thus produced, a swift vibratory movement or jar, which we know as an *earthquake*, runs through the earth's crust.

Extensive fissures are also formed, opening down to the regions where the rocks are liquid or plastic, and through these the melted rocks flow up to or toward the surface. That portion which flows out on the surface builds up a volcanic cone, while that which cools and solidifies below the surface, in the fissures, forms dikes. Thus among the igneous or eruptive rocks we have two great classes,—the *dike* rocks and the *volcanic* rocks.

It is important to observe that all these subterranean operations—the formation of continents, of mountain-ranges with all their attendant phenomena of folds, faults and cleavage, and every form and phase of earthquake and volcanic activity—depend upon or originate in the interior heat of the earth. And over against the *superficial* or *aqueous* agencies, originating in the *solar* heat and producing the *stratified* or *sedimentary* rocks, we set the *subterranean* or *igneous* agencies originating in the *central* heat, and producing the *unstratified* or *eruptive* rocks.

STRUCTURAL GEOLOGY.

In geology, just as in biology, there are two ways of studying structure,—the small way and the large way. In the case of an organism, we may select a single part or organ, and, disregarding its external form and relations to other parts, observe its composition and minute structure, the various forms and arrangements of the cells, etc. This is histology, and it is the complement of that larger method of studying structure which is ordinarily understood by anatomy.

The divisions of structural geology corresponding to histology and anatomy are *lithology* and *petrology*. Lithology is an in-door science; we use the microscope largely, and work with hand specimens or thin sections of the rocks, observing the composition and those small structural features which go under the general name of texture.

In petrology, on the other hand, we consider the larger kinds of rock-structure, such as stratification, jointing, folds, faults, cleavage, etc.; and it is essentially an out-door science, since to study it to the best advantage we must have, not hand specimens, but ledges, cliffs, railway-cuttings, gorges, and mountains.

LITHOLOGY.

A *rock* is any mineral, or mixture of minerals, occurring in masses of considerable size. This distinction of size is the only one that can be made between rocks and minerals, and that is very indefinite. A rock, whether composed of one mineral or several, is always an aggregate; and therefore no single crystal or mineral-grain can properly be called a rock.

Before proceeding to study particularly the various kinds of rocks, a little more preliminary work should be done. As already intimated, the more important characteristics of rocks may be grouped under two general heads,—*composition* and *texture*.

Composition of Rocks.

Rocks are properly defined as large masses or aggregates of mineral matter, consisting in some cases of one and in other cases of several mineral species. Hence it is clear that the composition of rocks is of two kinds: chemical and mineralogical; for the various chemical elements are first combined to form minerals, and then the minerals are combined to form rocks.

Of course those minerals and elements which can be described as principal or important rock-constituents must be the common minerals and elements. Now it is a very important and convenient fact that although chemists recognize about sixty-five elementary substances, and these are combined to form nearly one thousand mineral species, yet both the *common* elements and the *common* minerals are few in number.

So that, although it is very desirable and even necessary for the student of lithology to know something of chemistry and mineralogy, it by no means follows that he or she must be master of those sciences. A knowledge of the chemical and physical characteristics of a few common minerals is all that is absolutely essential, though it may be added that an excess of wisdom in these directions is no disadvantage.

Chemical Composition of Rocks.

The elementary substances of which rocks are chiefly composed, which make up the main mass of the earth so far as we are acquainted with it, number only fourteen:—

Non-Metallic or Acidic Elements.—Oxygen, silicon, carbon, sulphur, chlorine, phosphorus, and fluorine.

Metallic or Basic Elements.—Aluminum, magnesium, calcium, iron, sodium, potassium, and hydrogen.

The elements are named in each group in about the order of their relative abundance; and to give some idea of the enormous differences in this respect it may be stated that two of the elements—oxygen and silicon—form more than half of the earth's crust.

Silicon, calcium, and fluorine, although exceedingly abundant, are also very difficult to obtain in the free or uncombined state, and specimens large enough to exhibit to a class would be very expensive. With these exceptions, however, examples of these common rock-forming elements are easily obtained.

My purpose in calling attention to this point is simply to suggest that the proper way to begin the study of minerals and rocks with children is to first familiarize them with the elements of which they are composed. The most important thing to be known about any mineral is its chemical composition; and when a child is told that a mineral—corundum, for example—is composed of oxygen and aluminum, he should have a distinct conception of the properties of each of those elements, for otherwise corundum is for him a mere compound of names.

It is very important, too, if the pupil has not already studied chemistry, that he should be led to some comprehension of the nature of chemical union and of the difference between a chemical compound and a mechanical mixture. For this purpose a few simple experiments (the details of which would be out of place here) with the more common and familiar elements will be sufficient. Mrs. Richard's "First Lessons in Minerals" should be introduced here.

Mineralogical Composition of Rocks.

The fourteen elements named above are combined to form about fifty minerals with which the student of geology should be acquainted; but not more than one-half of these are of the first importance. It is desired to lay especial emphasis upon the importance of a perfect familiarity with these few common minerals. There is nothing else in the whole range of geology so easily acquired which is at the same time so valuable; for it is entirely impossible to comprehend the definitions of rocks, or to recognize rocks certainly and scientifically, unless we are acquainted with their constituent minerals.

With one or two exceptions, these common rock-forming minerals may be easily distinguished by their physical characters alone, so that their certain recognition is a matter of the simplest observation, and entirely within the capacity of young children. Furthermore, being common, specimens of these minerals are very easily obtained, so that there is no reason why teachers should not here adopt the best method and place a specimen of each mineral in the hands of each pupil. Typical examples, large enough to show the characteristics well, ought not to cost, on the average, over two cents apiece.

A MINERAL is an inorganic body having theoretically a definite chemical composition, and usually a regular geometric form.

THE PRINCIPAL CHARACTERISTICS OF MINERALS.—These may be grouped under the following general heads:—

(1) *Composition*, (2) *Crystalline form*, (3) *Hardness*, (4) *Specific gravity*, (5) *Lustre*, (6) *Color and Streak*.

1. *Composition.*—This, according to the definition of a mineral, ought to be *definite*, and expressible by a chemical formula. When it is not so, we usually consider that the mineral is partially decomposed, or that we are dealing with a mixture of minerals. It is well to impress upon the mind of the pupil the important fact that the more fundamental properties of the elements, such as specific gravity and lustre, are not lost when they combine, but may be traced in the compounds. In other words, the properties of minerals are, in a very large degree, the average of the properties of the elements of which they are composed; minerals in which heavy metallic elements predominate being heavy and metallic, and *vice versa*.

To fully appreciate this point it is only necessary to compare a mineral like galenite—a common ore of lead, and containing nearly 87 per cent. of that heavy metal; or hematite (specimen 13), containing 70 per cent. of another heavy metal, iron—with quartz (specimen 15), which is composed in nearly equal parts of oxygen and silicon, two typical non-metallic elements. Many minerals contain water, *i.e.*, are hydrated. Now water, whether we consider the liquid or solid state, is one of the lightest and softest of mineral constituents; and it is a very important fact that hydrated minerals are invariably lighter and usually softer than anhydrous species of otherwise similar composition. Other striking illustrations of this principle will be pointed out in the descriptions of the minerals which follow.

2. *Crystalline form.*—A crystal is bounded by plane surfaces symmetrically ar-

ranged with reference to certain imaginary lines passing through its centre and called axes. Crystals of the same species are always constant in the angles between like planes, while similar angles usually vary in different species; so that each species has its own peculiar form.

"Besides external symmetry of form, crystallization produces also regularity of internal structure, and often of fracture. This regularity of fracture, or tendency to break or cleave along certain planes, is called cleavage. The surface afforded by cleavage is often smooth and brilliant (see specimens 17, 18, and 21), and is always parallel with some external plane of the crystal. It should be understood that the cleavage lamellæ are not in any sense present before they are made to appear by fracture."—(Dana.)

Crystals are arranged in six systems, based upon the number and relations of the axes, as follows:—

Isometric System.—Three equal axes crossing at right angles. Example, cube.

Tetragonal System.—Two axes equal, third unequal, all crossing at right angles. Example, square prism.

Orthorhombic System.—Three unequal axes, but intersections all at right angles. Example, rhombic prism.

Monoclinic System.—Three unequal axes, one intersection oblique. Example, oblique rhombic prism.

Triclinic System.—Three unequal axes, all crossing obliquely. Example, oblique rhomboidal prism.

Hexagonal System.—Three equal axes lying in one plane and intersecting at angles of 60°, and a fourth axis crossing each of these at right angles and longer or shorter. Example, hexagonal prism.

By the truncation and bevelment of the angles and edges of these fundamental forms a vast variety of secondary forms are produced. The limits of the guide will not permit us to follow this topic farther; but it may be added that for the proper elucidation of even the simpler crystalline forms the teacher should be provided with a set of wooden crystal models and Dana's "Text-Book of Mineralogy."

The crystallization of a mineral may be manifested in two ways: first, by the regularity of its internal structure or molecular arrangement, as shown by cleavage and the polarization of transmitted light; and, second, by the regularity of external form which follows, *under favorable conditions*, as a necessary consequence of symmetry in the arrangement of the molecules.

When a mineral is entirely devoid of crystalline structure, both externally and internally, it is said to be *amorphous*.

Perfect and distinct crystals are the rare exception, most mineral specimens being simply aggregates of imperfect crystals. In such cases, and when the mineral is amorphous, the *structure* of the *mass* may be:—

Columnar or **fibrous**.

Lamellar, foliaceous, or **micaceous.**

Granular.—When the grains or crystalline particles are invisible to the naked eye the mineral is called *impalpable, compact,* or *massive.*

And the *external form* of the mass may be:—

Botryoidal, having grape-like surfaces.

Stalactitic, forming stalactites or pendant columns.

Amygdaloidal or **Concretionary,** forming separate globular masses in the enclosing rock.

Dendritic, branching or arborescent.

3. *Hardness.*—By the hardness of a mineral we mean the resistance which it offers to abrasion. But hardness is a purely relative term, calcite, for example, being hard compared with talc, but very soft compared with quartz. Hence mineralogists have found it necessary to select certain minerals to be used as a standard of comparison for all others, and known as the *scale of hardness.* These are arranged at nearly equal intervals all the way from the softest mineral to the hardest, as follows:—

Scale of Hardness.

1) Talc.	((4) Fluorite.	(8) Topaz or Beryl.
(2) Gypsum.	(5) Apatite.	(9) Corundum.
(3) Calcite.	(6) Orthoclase.	(10) Diamond.
	(7) Quartz.	

If a mineral scratches calcite and is scratched by fluorite, we say its hardness is between 3 and 4, perhaps 3.5; if it neither scratches nor is scratched by orthoclase, its hardness is 6; and so on. There are very few minerals harder than quartz, and hence the first seven members of the scale are sufficient for all ordinary purposes; and these are all included in the series of specimens accompanying this Guide.

Although it is desirable to be acquainted with the scale of hardness, and to understand how to use it, still the student will learn, after a little practice, that almost as good results may be obtained much more conveniently by the use of his thumb-nail and a good knife-blade or file. Talc and gypsum are easily scratched with the nail; calcite and fluorite yield easily to the knife or file, apatite with more difficulty; while orthoclase is near the limit of the hardness of ordinary steel, and quartz is entirely beyond it.

4. *Specific Gravity.*—The specific gravity of a mineral, by which we mean its weight as compared with the weight of an equal volume of water, is determined by weighing it first in air and then in water, and dividing the weight in air by the difference of the two weights. Minerals exhibit a wide range in specific gravity; from petroleum, which floats on water, to gold, which is nearly twenty times heavier than water. Although this is one of the most important properties of minerals, yet, being more difficult to measure than hardness, it is less valuable as an aid in distinguishing species. One can with practice, however, estimate the density

of a mineral pretty closely by lifting it in the hand.

5. *Lustre.*—Of all the properties of minerals depending on their relations to light the most important is lustre, by which we mean the quality of the light reflected by a mineral as determined by the character or minute structure of its surface. Two kinds of lustre, the *metallic* and *vitreous*, are of especial importance; in fact all other kinds are merely varieties of these.

The metallic lustre is the lustre of all true metals, as copper and tin, and characterizes nearly all minerals in which metallic elements predominate. The vitreous lustre is best exemplified in glass, but belongs to most minerals composed chiefly of non-metallic elements. Metallic minerals are always opaque, but vitreous minerals are often transparent.

Other kinds of lustre are the *adamantine* (the lustre of diamond), *resinous*, *pearly*, and *silky*. When a mineral has no lustre, like chalk, it is said to be dull.

It should be made clear to children that lustre and color are entirely distinct and independent. Thus, iron, copper, gold, silver, and lead are all metallic; while white or colorless quartz, black tourmaline, green beryl, red garnet, etc., are all vitreous. Generally speaking, any color may occur with any lustre.

6. *Color and Streak.*—The colors of minerals are of two kinds,—*essential* and *non-essential*. By the essential color in any case we mean the color of the mineral itself in its purest state. The non-essential colors, on the other hand, are chiefly the colors of the impurities contained in the minerals.

Metallic minerals, which are always opaque, usually have essential colors; but vitreous minerals, which are always more or less transparent, often have non-essential colors. The explanation is this: In opaque minerals we can only see the impurities immediately on the surface, and these are, as a rule, not enough to affect its color; but in diaphanous minerals we look *into* the specimen and see impurities below the surface, and thus bring into view, in many cases, sufficient impurity so that its color drowns that of the mineral.

To prove this we have only to take any mineral (serpentine is a good example in our series) having a non-essential color, and make it opaque by pulverizing it or abrading its surface, when the non-essential color, the color of the impurity, immediately disappears; just as water, yellow with suspended clay, becomes white when whipped into foam, and thus made opaque.

What we understand by the *streak* of a mineral is its essential color, the color of its powder; and it is so called because the powder is most readily observed by scratching the surface of the mineral, and thereby pulverizing a minute portion of it. The streak and hardness are thus determined at the same time. The streak of soft minerals is easily determined by rubbing them on any white surface of suitable hardness, as paper, porcelain, or Arkansas stone.

Essential and Accessory Minerals.—Lithologists, regarding minerals as constituents of rocks, divide them into two great classes: the *essential* and the *accessory*. The essential constituents of a rock are those minerals which are essential to the definition of the rock. For example, we cannot properly define granite without naming quartz and orthoclase; hence these are essential constituents of granite; and if either of these minerals were removed from granite it would not be granite any longer, but some other rock. But other minerals, like tourmaline and garnet, may be indifferently present or absent; it is granite still; hence they are merely accidental or accessory constituents. They determine the different *varieties* of granite, while the essential minerals make the *species*.

This classification, of course, is not absolute, for in many cases the same mineral forms an essential constituent of one rock and an accessory constituent of another. Thus, quartz is essential in granite, but accessory in diorite.

PRINCIPAL MINERALS CONSTITUTING ROCKS.—Having studied in a general way the more important characteristics of minerals, brief descriptions of the chief rock-forming species are next in order. We will notice first and principally those minerals occurring chiefly as *essential* constituents of rocks.

1. **Graphite.**—Essentially pure carbon, though often mixed with a little iron oxide. Crystallizes in hexagonal system, but usually foliated, granular, or massive. Hardness, 1-2, being easily scratched with the nail. Sp. gr., 2.1-2.3. Lustre, metallic; an exception to the rule that acidic elements have non-metallic or vitreous lustres. Streak, black and shining (see pencil-mark on white paper). Color, iron-black. Slippery or greasy feel. Every *black-lead* pencil is a specimen of graphite. Specimen 9.

The different kinds of mineral coal are, geologically, as we have seen, closely related to graphite, but they are such familiar substances that they need not be described here.

2. **Halite** (common salt).—Chloride of sodium: chlorine, 60.7; sodium, 39.3; = 100. Isometric system, usually forming cubes. Hardness, 2.5, a little harder than the nail. Sp. gr., 2.1-2.6. Lustre, vitreous. Streak and color both white, and hence color is essential. Often transparent. Soluble; taste, purely saline. In specific gravity and lustre it is a good example of a mineral in which an acidic element predominates. Specimen 11.

3. **Limonite.**—Hydrous sesquioxide of iron: oxygen, 25; iron, 60; water, 15; = 100. Usually amorphous; occurring in stalactitic and botryoidal forms, having a fibrous structure; and also concretionary, massive, and earthy (yellow ochre). Hardness, 5-5.5. Sp. gr., 3.6-4. Lustre, vitreous or silky, inclining to metallic, and sometimes dull. Color, various shades of black, brown, and yellow. Streak, ochre-yellow; hence color partly non-essential. Specimen 12.

4. **Hematite.**—Sesquioxide of iron: oxygen, 30; iron, 70; = 100. Hexagonal system, in distinct crystals, but usually lamellar, granular, or compact,—columnar, botryoidal, and stalactitic forms being common. Hardness, 5.5-6.5; good crystals are harder than steel. Sp. gr., 4.5-5.3. Lustre, metallic, sometimes dull. Color, iron-black, but red when earthy or pulverized (red ochre). Streak, red, and color, therefore, mainly non-essential; sometimes attracted by the magnet. Specimen 13.

Hematite has the same composition as limonite, minus the water; and by comparing the hardness and specific gravity of these two minerals we see that they are a good illustration of the principle that hydrous minerals are softer and lighter than anhydrous minerals of analogous composition. Limonite and hematite are two great natural coloring agents, and almost all yellow, brown, and red colors in rocks and soils are due to their presence.

5. **Magnetite.**—Protoxide and sesquioxide of iron: oxygen, 27.6; iron, 72.4; = 100. Isometric system, usually in octahedrons or dodecahedrons. Most abundant variety is coarsely to finely granular, sometimes dendritic. Hardness, 5.5-6.5, same as hematite. Sp. gr., 4.9-5.2. Lustre, metallic. Color and streak, iron-black, and hence color essential. Strongly magnetic; some specimens have distinct polarity, and are called loadstones. Specimen 14.

The three iron-oxides just described—limonite, hematite, and magnetite—are all important ores of iron, and form a well-marked natural series. Thus limonite is never, hematite is usually, and magnetite is always, crystalline. Again, limonite with 60 per cent. of iron is never magnetic, hematite with 70 per cent. is sometimes magnetic, while magnetite with 72.4 per cent. is always magnetic. As the iron increases so does the magnetism. We have here an excellent illustration of the principle that the properties of the elements can be traced in those minerals in which they predominate. Iron is the only strongly magnetic element: magnetite contains more iron than any other mineral, and it is the only strongly magnetic mineral.

These three iron-ores are easily distinguished from each other by the color of their powders or streak,—limonite yellow, hematite red, and magnetite black,—and from all other common minerals by their high specific gravity.

6. **Quartz.**—Oxide of silicon or silica: oxygen, 53.33; silicon, 46.67; = 100. Hexagonal system. The most common form is a hexagonal prism terminated by a hexagonal pyramid. Also coarsely and finely granular to perfectly compact, like flint; the compact or cryptocrystalline varieties often assuming botryoidal, stalactitic, and concretionary forms. It has no cleavage, but usually breaks with an irregular, conchoidal fracture like glass. Hardness, 7, being No. 7 of the scale; scratches glass easily. Sp. gr., 2.5-2.8. Lustre, vitreous. Pure quartz is colorless or white, but by admixture of impurities it may be of almost any color. Streak always white or light colored. Quartz is usually, as in specimen 15, transparent and glassy, but may be translucent or opaque. It is almost absolutely infusible and insoluble.

The varieties of quartz are very numerous, but they may be arranged in two great groups:—

1. *Phenocrystalline* or *vitreous* varieties, including rock-crystal, amethyst, rose quartz, yellow quartz, smoky quartz, milky quartz, ferruginous quartz, etc.

2. *Cryptocrystalline* or *compact* varieties, including chalcedony, carnelian, agate, onyx, jasper, flint, chert, etc. Only three varieties, however, are of any great geological importance; these are: common glassy quartz (spec. 15), flint (spec. 16), and chert.

Quartz is one of the most important constituents of the earth's crust, and it is also the hardest and most durable of all common minerals. We have already observed (p. 12) that it is entirely unaltered by exposure to the weather; *i.e.*, it cannot be decomposed; and, being very hard, the same mechanical wear which, assisted by more or less chemical decomposition, reduces softer minerals to an impalpable powder or clay, must leave the quartz chiefly in the form of sand and gravel. This agrees with our observation that sand (spec. 30), especially, is usually merely pulverized quartz.

Opal is a mineral closely allied to quartz, and may be mentioned in this connection. It is of similar composition, but contains from 5 to 20 per cent. of water, and is decidedly softer and lighter. Hardness, 5.5-6.5; sp. gr., 1.9-2.3.

7. **Gypsum.**—Hydrous sulphate of calcium: sulphur trioxide (SO_3), 46.5; lime (CaO), 32.6; water (H_2O), 20.9; = 100. Monoclinic system. Often in distinct rhombic crystals; also foliated, fibrous, and finely granular. Hardness, 1.5-2; the hardest varieties being No. 2 of the scale of hardness. Sp. gr., 2.3. Lustre, pearly, vitreous, or dull. Color and streak usually white or gray. The principal varieties of gypsum are (*a*) *selenite*, which includes all distinctly crystallized or transparent gypsum; (*b*) *fibrous gypsum* or *satin-spar*; (*c*) *alabaster*, fine-grained, light-colored, and translucent. Gypsum is easily distinguished from all common minerals resembling it by its softness and the fact that it is not affected by acids. Specimen 17.

8. **Calcite.**—Carbonate of calcium: carbon dioxide (CO_2), 44; lime (CaO), 56; = 100. Hexagonal system, usually in rhombohedrons, scalenohedrons, or hexagonal prisms. Cleavage rhombohedral and highly perfect (specimen 18). Also fibrous and compact to coarsely granular, in stalactitic, concretionary, and other forms. Hardness, 2.5-3.5, usually 3 (see scale of hardness). Sp. gr., 2.5-2.75. Lustre, vitreous. Color and streak usually white. Transparent crystallized calcite is known as *Iceland-spar*, and is remarkable for its strong double refraction. When finely fibrous it makes a *satin-spar* similar to gypsum. Geologically speaking, calcite is a mineral of the first importance, being the sole essential constituent of all limestones. It is readily distinguished from allied species by its perfect rhombohedral cleavage; by its softness, being easily scratched with a knife; and above all by its lively effervescence with acids, for it is the *only common* mineral effervescing *freely* with *cold dilute* acid. To apply this test it is only necessary to touch the specimen with a drop of dilute chlorohydric acid. The effervescence, of course, is due to the escape of the carbon dioxide in a gaseous form. Specimen 18.

9. **Dolomite.**—Carbonate of calcium and magnesium: carbonate of calcium ($CaCO_3$), 54.35; carbonate of magnesium ($MgCO_3$), 45.65; = 100. Hexagonal system, being nearly isomorphous with calcite. Rhombohedral cleavage perfect. Hardness, 3.5-4; sp. gr., 2.8-2.9, being harder and heavier than calcite. Lustre, color, and streak same as for calcite, from which it is most easily distinguished by its non-effervescence or only feeble effervescence with cold dilute acid, though effervescing freely with strong or hot acid. Spec. 19.

10. **Siderite.**—Carbonate of iron: carbon dioxide (CO_2), 37.9; protoxide of iron (FeO), 62.1; = 100. Crystallization and cleavage essentially the same as for calcite and dolomite. Hardness, 3.5-4.5, and sp. gr., 3.7-3.9. Lustre, vitreous. Color, white, gray, and brown. Streak, white. With acid, siderite behaves like dolomite. It is distinguished from both calcite and dolomite by its high specific gravity, which is easily explained by the fact that it is largely composed of the heavy element, iron.

With one exception, the fifteen minerals which we have yet to study belong to the class of silicates, which includes more than one-fourth of the known species of minerals, and, omitting quartz and calcite, all of the really important rock-constituents. The silicate minerals may be very conveniently divided into two great groups, the *basic* and *acidic*. This is not a sharp division; on the contrary, there is a perfectly gradual passage from one group to the other; and yet this is, for geological purposes at least, a very natural classification. The dividing line falls in the neighborhood of 60 per cent. of silica; *i.e.*, all species containing this proportion of silica or *less* are classed as basic, since in them the basic elements predominate; while those containing *more* than 60 per cent. of silica are classed as acidic, because their characteristics are determined chiefly by the acid element or silica. The principal bases occurring in the silicates, named in the order of their relative importance, are aluminum, magnesium, calcium, iron, sodium, and potassium; and of these, magnesium, calcium, iron, and usually sodium, are especially characteristic of basic species.

Iron is the heaviest base; but all the bases, except sodium and potassium, are heavier than the acid—silica; consequently basic minerals must be, as a rule, heavier than acidic minerals. And since basic minerals contain more iron than acidic, they must be darker colored. In general, we say, *dark, heavy* silicates are *basic*, and *vice versa*. All this is of especial importance because in the rocks nature keeps these two classes separate in a great degree.

11. **Amphibole.**—Silicate of aluminum, magnesium, calcium, iron, and sodium. The bases occur in very various proportions, forming many varieties; but the only variety of especial geological interest is *hornblende*, the average percentage composition of which is as follows: silica (SiO_2), 50; alumina (Al_2O_3), 10; magnesia (MgO), 18; lime (CaO), 12; iron oxide (FeO and Fe_2O_3), 8; and soda (Na_2O), 2; = 100. Monoclinic system: usually in rhombic or six-sided prisms which may be short and thick, but are more often acicular or bladed. Hardness, 5-6; sp. gr., 2.9-3.4. Lustre, vitreous; color, black and greenish black; and streak similar to color, but much paler. Compare with quartz, and observe the strong contrast in color possible with minerals having the same lustre. Specimen 20.

12. **Pyroxene.**—Like amphibole, this species embraces many varieties, and these exhibit a wide range in composition; but of these *augite* alone is an important rock-constituent. Hence in lithology we practically substitute for amphibole and pyroxene, hornblende, and augite respectively.

Augite is very similar in composition to hornblende, but contains usually more lime and less alumina and alkali. Physically, too, these minerals are almost identical, crystallizing in the same system and in very similar forms, and agreeing in hardness, color, lustre, and streak. Augite is heavier than hornblende, sp. gr., 3.2-3.5. A certain prismatic angle, which in augite is 87°5′, is 124°30′ in hornblende. Slender, bladed crystals are more common with hornblende than augite. When examined in thin sections with the polarizer, augite does not afford the phenomenon of dichroism, which is strongly marked in hornblende. However, as these minerals commonly occur in the rocks, in small and imperfect crystals, these distinctions can only be observed in thin sections under the microscope; so that, as regards the naked eye, they are practically indistinguishable.

It might appear at first that the distinction of minerals so nearly identical is not an important matter; but nature has decreed otherwise. Augite and hornblende are typical examples of basic minerals; but augite is, both in its composition and associations, the more basic of the two. In proof of this we need only to know that it very rarely occurs in the same rock with quartz, while hornblende is found very commonly in that association. Quartz in a rock means an excess of acid or silica, and almost necessarily implies the absence of highly basic minerals. In other words, hornblende is often, and augite very rarely, found in connection with acidic minerals; and it is this difference of association chiefly that makes their distinction essential to the proper recognition of rocks; while at the same time it affords an easy, though of course not absolutely certain, means of determining whether the black constituent of any particular rock is hornblende or augite.

Mica Family.—Mica is not the name of a single mineral, but of a whole family of minerals, including some half-dozen species. Only two, however,—muscovite and biotite,—are sufficiently abundant to engage our attention. These are complex, basic silicates of aluminum, magnesium, iron, potassium, and sodium. The crystallization of biotite is hexagonal, and of muscovite monoclinic; but both occur commonly in flat six-sided forms. Undoubtedly the most important and striking characteristic of the whole mica family is the remarkably perfect cleavage parallel with the basal planes of the crystals, and the wonderful *thinness*, and above all the *elasticity*, of the cleavage lamellæ. The cleavage contrasts the micas with all other common minerals, and makes their certain identification one of the easiest things in lithology. The micas are soft minerals, the hardness ranging from 2 to 3, and being usually easily scratched with the nail. Sp. gr. varies from 2.7-3.1. Lustre, pearly; and streak, white or uncolored.

The distinguishing features of muscovite and biotite are as follows:—

13. **Muscovite.**—Contains 47 per cent. of silica, 3 per cent. of sesquioxide of iron, and 10 per cent. of alkalies, chiefly potash; and the characteristic colors are white, gray, and, more rarely, brown and yellow. Non-dichroic. Usually found in association with acidic minerals. The mica used in the arts is muscovite. Specimen 21.

14. **Biotite.**—Contains only 36 per cent. of silica, 20 per cent. of oxide of iron, and 17 per cent. of magnesia; colors, deep black to green. Strongly dichroic. Commonly occurs with other basic minerals. Compare color with per cent. of iron.

These differences are tabulated below:—

Muscovite =	*Biotite =*
Acidic mica.	Basic mica.
Non-ferruginous mica.	Ferruginous mica.
Potash mica.	Magnesian mica.
White mica.	Black mica.
Non-dichroic mica.	Dichroic mica.

Feldspar Family.—Like mica, *feldspar* is the name of a family of minerals; and these are, geologically, the most important of all minerals. They are, above all others, the minerals of which rocks are made, and their abundance is well expressed in the name,—feldspar being simply the German for field-spar, implying that it is the common spar or mineral of the fields.

Chemically, the feldspars are silicates of aluminum and potassium, sodium or calcium. They crystallize in the monoclinic and triclinic systems; and all possess easy cleavage in two directions at right angles to each other, or nearly so. The general physical characters, including the cleavage, are well exhibited in the common species, orthoclase (specimen 22).

In hardness the feldspars range from 5 to 7, being usually near 6, and almost always distinctly softer than quartz. Sp. gr. varies from 2.5-2.75; lustre, from vitreous to pearly; color, from white and gray to red, brown, green, etc., but usually light. Streak, always white; rarely transparent. By exposure to the weather, feldspars gradually lose their alkalies and lime, become hydrated, and are changed to kaolin or common clay. A similar change takes place with the micas, augite, and hornblende; but these species, being usually rich in iron, make clays which are much·darker colored than those derived from feldspars. The fact that the feldspars contain little or no iron undoubtedly explains their low specific gravity and light colors, as compared with the other minerals just named. The only common minerals for which the feldspars are liable to be mistaken are quartz and the carbonates. From the latter they are easily distinguished by their superior hardness and non-effervescence with acids; and from the former, by possessing distinct cleavage, by being rarely transparent, by being somewhat softer, and by changing to clay on exposure to the weather.

The feldspars of greatest geological interest are five in number, and may be classified chemically as follows:—

> Orthoclase,—silicate of aluminum and potassium, or potash feldspar.
>
> Albite,—silicate of aluminum and sodium, or soda feldspar.
>
> Anorthite,—silicate of aluminum and calcium, or lime feldspar.
>
> Oligoclase,—silicate of aluminum, sodium, and calcium, or soda-lime feldspar.
>
> Labradorite,—silicate of aluminum, calcium, and sodium, or lime-soda feldspar.

This appears like a complex arrangement, but it can be simplified. Orthoclase crystallizes in the monoclinic system, and all the other feldspars in the triclinic system. With the exception of albite, which is a comparatively rare species, the triclinic feldspars all contain less silica than orthoclase; *i.e.*, are more basic. This is shown by the subjoined table giving the average composition of each of the feldspars:—

	SiO_2	Al_2O_3	K_2O	Na_2O	CaO		Total.
Orthoclase,	65	18	17	--	--	=	100

Albite,	68	20	--	12	--	=	100
Oligoclase,	62	24	--	9	5	=	100
Labradorite,	53	30	--	4	13	=	100
Anorthite,	43	37	--	--	20	=	100

As we should naturally expect, the triclinic feldspars occur usually with other basic minerals, while the monoclinic species, orthoclase, is acidic in its associations; furthermore, the triclinic feldspars are often intimately associated with each other, but are rarely important constituents of rocks containing much orthoclase. In other words, the distinction of orthoclase from the basic or triclinic feldspars is important and comparatively easy, while the distinction of the different basic feldspars from each other is both unimportant and difficult. Hence, in lithology, we find it best to put all these basic feldspars together, as if they were one species, under the name *plagioclase*, which refers to the oblique cleavage of all these feldspars, and contrasts with *orthoclase*, which refers to the right-angled cleavage of that species.

This statement of the relations of the feldspars is, of course, beyond the comprehension of many children, and yet it should be understood by the teacher who would lead the children to any but the most superficial views.

15. **Orthoclase.**—This is the common feldspar, and the most abundant of all minerals, being the principal constituent of granite, gneiss, and many other important rocks. The most characteristic colors are white, gray, pinkish, and flesh-red. Specimen 22.

16. **Plagioclase.**—Like orthoclase, these species may be of almost any color; yet these two great divisions of the feldspars are usually contrasted in this respect. Thus, bluish and grayish colors are most common with plagioclase, and white or reddish colors with orthoclase. Specimen 23 is labradorite, and, in every respect, a typical example of plagioclase. On certain faces and cleavage-surfaces of the plagioclase crystals we may often observe a series of straight parallel lines or bands which are often very fine,—fifty to a hundred in a single crystal. These striæ are due to the mode of twinning, and are of especial importance, since, while they are very characteristic of plagioclase, they never occur in orthoclase. As stated, these twinning striæ in plagioclase are often visible to the naked eye; and when they are not, they may usually be revealed by examining a thin section under the microscope with polarized light. Plagioclase decays much more rapidly when exposed to the weather than orthoclase. This point becomes perfectly clear when we compare weathered ledges of diabase (or any trap-rock, see specimen 2) and granite; for plagioclase is the principal constituent of the former rock, and orthoclase of the latter.

Hydrous Silicates.—Many silicates contain water, and some of these are of great geological importance. What has been stated on a preceding page concerning the softness and lightness of hydrated minerals is especially applicable here; for all the geologically important hydrous silicates are distinctly softer and lighter than anhydrous minerals of otherwise similar composition. Furthermore, they usually have an unctuous or slippery feel; and, with one exception (kaolin), are of a green or greenish color.

17. **Kaolinite (Kaolin).**—Hydrous silicate of aluminum: silica (SiO_2), 46; alumina (Al_2O_3), 40; and water (H_2O), 14; = 100. Orthorhombic system, in rhombic or hexagonal scales or plates, but usually earthy or clay-like. Hardness, 1-2.5; sp. gr., 2.4-2.65. The pure mineral is white; but it is usually colored by impurities, the principal of which are iron oxides and carbonaceous matter. Kaolin is the most abundant of all the hydrous silicates, and it is the basis and often the sole constituent of common clay,—a very common mineral, but rarely pure. We have already (p. 11) noticed the mode of origin of kaolin or clay. It results from the decomposition of various aluminous silicate minerals, especially the feldspars. Under the combined influence of carbon dioxide and moisture, feldspars give up their potassium, sodium, and calcium, and take on water, and the result is kaolin. This mineral is believed to be always a decomposition product. Perhaps the best, or at least the most convenient, test for kaolin is the argillaceous odor, the odor of moistened clay. Specimen 24.

18. **Talc.**—Hydrous silicate of magnesium: silica (SiO_2), 63 (acidic); magnesia (MgO), 32; water (H_2O), 5; = 100. Orthorhombic system, but rarely in distinct crystals. Cleavage in one direction very perfect; the cleavage lamellæ are flexible, but not elastic, as in mica. Hardness, 1; see scale. Sp. gr., 2.55-2.8. Lustre, pearly. Color, apple-green to white; and streak, white. The feel is very smooth and greasy; and, in connection with the color and foliation, affords the best means of distinguishing talc from allied minerals. Talc sometimes results from the alteration of augite, hornblende, and other minerals, but it is not always nor usually an alteration product.

19. **Serpentine.**—Hydrous silicate of magnesium: silica (SiO_2), 44 (basic); magnesia (MgO), 44; water (H_2O), 12; = 100. Essentially amorphous. Hardness, 2.5-4; sp. gr., 2.5-2.65. Lustre, greasy, waxy, or earthy. Color, various shades of green and usually darker than talc, but streak always white. Feel, smooth, sometimes greasy. Distinguished from talc by its hardness, compactness, and darker green. Sometimes results from the alteration of olivine and other magnesian minerals, but usually we are to regard it as an original mineral. Specimen 25.

20. **Chlorite.**—This is, properly, the name of a group of highly basic minerals of very variable composition, but they are all essentially hydrous silicates of aluminum, magnesium, and iron; and the average composition of the most abundant species, prochlorite, is as follows: silica (SiO_2), 30; alumina (Al_2O_3), 18; magnesia (MgO), 15; protoxide of iron (FeO), 26; and water (H_2O), 11; = 100. The chlorites crystallize in several different systems, but in all there is a highly perfect cleavage in one direction, giving, as in talc, a foliated structure with flexible but inelastic laminæ. The cleavage scales, however, are sometimes minute, and the structure massive or granular. Hardness of prochlorite, 1-2; between talc and serpentine. Sp. gr., 2.78-2.96. All the chlorites have a pearly to vitreous lustre. Color usually some shade of green; in prochlorite a dark or blackish green, darker than serpentine, as that is darker than talc. Streak, a lighter, whitish green. Less unctuous than talc, but more so than serpentine. The chlorites are produced very commonly, but

not generally, by the alteration of basic anhydrous silicates, like augite and hornblende. Specimen 26.

21. **Hydro-mica.**—This, too, is properly the name of a group of minerals; but for geological purposes they may be regarded as one species. Taking a general view of the composition, these are simply the anhydrous or ordinary micas, which we have already studied, with from 5 to 10 per cent. of water added. In crystallization and structure they are essentially mica-like. Although not distinctly softer than the common micas, they are lighter, always more unctuous and slippery, and usually of a greenish color. The micaceous structure with *elastic* laminæ serves to distinguish the hydro-micas from other hydrous silicates.

22. **Glauconite.**—Hydrous silicate of aluminum, iron, and potassium: silica (SiO_2), 50; alumina (Al_2O_3), protoxide of iron (FeO), and potash (K_2O), together, 41; and water (H_2O), 9; = 100. Amorphous, forming rounded and generally loose grains, which often have a microscopic organic nucleus. It is dull and earthy, like chalk, and always soft, green, and light, but not particularly unctuous. Glauconite is the principal, often the sole, constituent of the rock greensand, which occurs abundantly in the newer geological formations, and is now forming in the deep water of the Gulf of Mexico and along our Atlantic sea-board. Specimen 27.

This completes our list of minerals occurring chiefly as *essential* constituents of rocks; and following are three of the more common and important minerals occurring chiefly as *accessory*, rarely as essential, rock-constituents.

23. **Chrysolite (Olivine).**—Silicate of magnesium and iron: silica (SiO_2), 41; magnesia (MgO), 51; protoxide of iron (Fe_2O_3), 8; = 100. Orthorhombic system; but usually in irregular glassy grains. Hardness, 6-7. Sp. gr., 3.3-3.5. Lustre, vitreous; color, usually some shade of green; and streak, white. Chrysolite sometimes closely resembles quartz, but its green color usually suffices to distinguish it. It is a common constituent of basalt and allied rocks. By absorption of water it is changed into serpentine and talc. See examples in specimen.

24. **Garnet.**—The composition of this mineral is extremely variable; but the most important variety is a basic silicate of aluminum and iron: silica (SiO_2), 37; alumina (Al_2O_3), 20; and protoxide of iron (FeO), 43; = 100. Isometric system, usually in distinct crystals, twelve-sided (dodecahedrons) and twenty-four-sided (trapezohedrons) forms being most common. Hardness, 6.5-7.5; average as hard as quartz. Sp. gr., 3.15-4.3; compare with the high percentage of iron. Lustre, vitreous; colors, various, usually some shade of red or brown; and streak, white. Some varieties contain iron enough to make them magnetic. Garnet is easily distinguished by its form, color, and hardness from all other minerals. It is a common but not an abundant mineral, occurring most frequently in gneiss, mica schist, and other stratified crystalline rocks. See examples in specimen.

25. **Pyrite.**—Sulphide of iron: sulphur, 53.3; iron, 46.7; = 100. Isometric system, occurring usually in distinct crystals, the cube and the twelve-sided form known as the pyritohedron being the most common. Hardness, 6-6.5, striking fire with steel. Sp. gr., 4.8-5.2; heavy because rich in iron. Lustre, metallic and splendent. Color, pale, brass-yellow, and streak, greenish or brownish. Pyrite is sometimes

mistaken for gold, but it is not malleable; while its color, hardness, and specific gravity, combined, easily distinguish it from all common minerals. As an accessory rock-constituent, pyrite occurs usually in isolated cubes or pyritohedrons. Specimen 10.

Textures of Rocks.

Texture is a general name for those smaller structural features of rocks which can be studied in *hand specimens*, and which depend upon the *forms* and *sizes* of the *constituent particles* of the rocks, and the *ways* in which these are *united*.

By "constituent particles" we mean, not the atoms or molecules of matter composing the rocks, but the *pebbles* in conglomerate, *grains of sand* in sandstone, *crystals of quartz, feldspar,* and *mica* in granite, etc. The four most important textures are:—

(1) *Fragmental texture.*—The rock is composed of mere irregular, angular, or rounded, but visible, fragments. Examples: sand, sandstone, gravel, conglomerate, etc. Specimens 30, 31, 28, 29.

(2) *Crystalline texture.*—The constituent particles are chiefly, at least, distinctly crystalline, as shown either by external form, or cleavage, or both. Examples: granite, diabase, gneiss, etc. Specimens 45, 1, 41.

(3) *Compact texture.*—The constituent particles are indistinguishable by the naked eye, but become visible under the microscope, appearing as separate crystalline grains or as irregular fragments. In other words, if, in the case of either the granular or crystalline textures, we conceive the particles to become microscopically small, then we have the compact texture. Examples: clay, slate, many limestones, basalt, etc. Specimens 34, 35, 39.

(4) *Vitreous texture.*—The texture of glass, in which the constituent particles are absolutely invisible even with the highest powers of the microscope, and may be nothing more than the *molecules* of the substance, which thus, so far as our powers of observation are concerned, presents a perfectly continuous surface. Examples: obsidian, glassy quartz, and some kinds of coal. Specimens 47, 15.

These four textures, which, it will be observed, are determined by the forms and sizes of the constituent particles, may be called the *primary* textures, because every rock *must* possess one of them. We cannot conceive of a rock which is neither fragmental, crystalline, compact, nor vitreous. But in addition to one of the primary textures, a rock may or may not have one or more of what may be called *secondary* textures. These are determined by the way in which the particles are united, the mode or pattern of the arrangement, etc. Following are definitions of the principal secondary textures:—

(1) *Laminated texture.*—This exists where the particles are arranged in thin, parallel layers, which may be marked simply by planes of division, or the alternate layers may be composed of particles differing in composition, form, size, or color, etc. Among the laminated textures we thus distinguish: (*a*) the *banded* texture, where the layers are contrasted in color, texture, or composition, but cohere, so

that there is no cleavage or easy splitting parallel with the stratification; and (*b*) the *schistose* or *shaly* texture, where such fissility or stratification-cleavage exists. If a fragmental, compact, or vitreous rock is fissile, we use the term *shaly*; but a fissile, crystalline rock is described as *schistose*. The banded texture may occur with the fragmental,—banded sandstones, etc.; with the crystalline,—many gneisses, etc. (specimen 41); with the compact,—many slates, limestones, felsites, etc. (specimens 34, 42); with the vitreous,—banded obsidian, furnace slags, and some coal. The schistose texture may occur with the crystalline,—mica schist, etc. (specimen 43); and the shaly texture with the compact and fragmental, but rarely with the vitreous.

(2) *Porphyritic texture.*—We have this texture when *separate* and *distinct crystals* of *any* mineral, but most commonly of feldspar, are enclosed in a *relatively* fine-grained base or matrix, which may be either crystalline, compact, or vitreous, but rarely fragmental. Specimens 5, 6, 7 are examples of the porphyritic compact texture.

(3) *Concretionary texture.*—When one or more constituents of a rock have the form, in whole or in part, not of distinct angular crystals, but of rounded concretions, the texture is described as concretionary, the concretions taking the place in this texture of the isolated crystals in the porphyritic texture. This texture occurs in connection with all the primary textures, but the most familiar example is oölitic limestone.

(4) *Vesicular texture.*—A rock has this texture when it contains numerous small cavities or vesicles. These are most commonly produced by the expansion of steam and other vapors when the rock is in a plastic state; and hence the vesicular texture is found chiefly in volcanic rocks. Except rarely, it is associated only with the compact texture,—ordinary stony lavas (specimen 49); and with the vitreous texture,—pumice (specimen 48).

(5) *Amygdaloidal texture.*—In the course of time the vesicles of common lava are often filled with various minerals deposited by infiltrating waters, giving rise to the amygdaloidal texture, from the Latin *amygdalum*, an almond, in allusion to a common form of the vesicles, or amygdules, as they are called, after being filled. The amygdaloidal texture is thus necessarily preceded by the vesicular, and is limited to the same classes of rocks. Specimen 50.

Besides the foregoing, there are many minor secondary textures. The rocks known as tufas have what may be called the *tufaceous* texture. Then we have kinds of texture depending on the *strength* of the union of the particles, as *strong, weak, friable, earthy,* etc.

Classification of Rocks.

			Unconsolidated.	Consolidated.
Sedimentary or Stratified Rocks. — **MECHANICALLY FORMED.**	*Conglomerate group.*		Gravel.	Conglomerate.
	Arenaceous group.		Sand.	Sandstone.
	Argillaceous group.		Clay.	Slate.

Sedimentary or Stratified Rocks. — CHEMICALLY AND ORGANICALLY FORMED.

Coal group.	*Iron-ore group.*	*Calcareous group.*	*Metamorphic group (Silicates).*	
			Acidic.	Basic.
			85 80 70 60	50 40 30
Peat. Lignite. Bit. Coal. Anthracite. Graphite. Asphaltum.	Limonite. Hematite. Magnetite. Siderite.	Limestone. Dolomite. Gypsum. Rock-salt. Phosphate Rock.	**Feldspathic.** Gneiss. / Syenite.	Diorite. / Norite.
	Siliceous group. Tripolite. Flint. Siliceous Tufa. Novaculite.		**Non-Feldspathic.** Mica Schist. / Hornbl. Schist. / Talc Schist. /	Amphibolite. / Chl. Schist. / Greensand, Serpentine.

Eruptive or Unstratified Rocks.

PLUTONIC.

This part of the classification is a blank, for the reason that no eruptive rocks are known which are chiefly composed of minerals belonging to the classes of Native Elements, Chlorides, Oxides, Sulphates, or Carbonates; *i.e.*, all eruptive rocks, so far as known, are principally composed of minerals belonging to the class of Silicates.

	Acidic	Basic
Feldspathic.	Granite. / Syenite.	Diorite. / Diabase.

VOLCANIC.

	Acidic	Basic
Feldspathic.	Rhyolite. / Obsidian. / Petrosilex.	Andesite. / Trachyte. Basalt. / Tachylite. / Porphyrite. / Felsite. Melaphyr

Classification of Rocks.

Having finished our preliminary observations on the characteristics of rocks, we are now about ready to begin a systematic study of the rocks themselves; but it is needful first to say a few words about the classification of rocks, since upon this depends not only the order in which we shall take the rocks up, but also the ideas that will be imparted concerning their relations and affinities. The classifications which have been proposed at different times are almost as numerous as the rocks themselves. Some of these are confessedly, and even designedly, artificial, as when we classify stones according to their uses in the arts, etc. But we want something more scientific, a *natural* classification; that is, one based upon the natural and permanent characteristics of rocks. Rocks have been classified according to chemical composition, mineralogical composition, texture, color, density, hardness, etc.; but these arrangements, taken singly or all combined, are inadequate.

A *natural* classification may be defined as a concise and systematic statement of the natural relations existing among the objects classified. Now the most important relations existing among rocks are those due to their different origins. We must not forget that lithology is a branch of geology, and that geology is first of all a *dynamical* science. The most important question that can be asked about any rock is, not What is it made of? but *How* was it made? What were the general forces or agencies concerned in its formation? Rocks are the material in which the earth's history is written, and what we want to know first concerning any rock is what it can tell us of the condition of that part of the earth at the time it was made and subsequently.

The geological agencies, as we have already learned, may be arranged in two great classes: first, the aqueous or superficial agencies originating in the solar heat, and producing the sedimentary or stratified rocks; and, second, the igneous or subterranean agencies originating in the central or interior heat, and producing the eruptive or unstratified rocks. Hence, we want to know first of any rock whether it is of aqueous or igneous origin. Then, if it is a sedimentary rock, whether it has been formed by the action chiefly of mechanical forces, or of chemical and organic forces. And, if it is an eruptive rock, whether it has cooled and solidified below the earth's surface in a fissure, and is a dike or trappean rock, or has flowed out on the surface and cooled in contact with the air, and thus become an ordinary lava or volcanic rock.

Here we have the outlines of our classification, and it will be observed that we have simply reached the conclusion, in a somewhat roundabout manner, that there should always be a general correspondence between the classification of rocks and the classification of the forces that produce them. The general plan of the preceding scheme of the classification must now be clear, and the details will be explained as we go along.

Descriptions of Rocks.

1.—Sedimentary or Stratified Rocks.

1. Mechanically formed or Fragmental Rocks.—These consist of materials

deposited from *suspension* in water, and the process of their formation is throughout chiefly mechanical. The materials deposited are mere fragments of older rocks; and, if the fragments are large, we call the newly deposited sediment gravel; if finer, sand; and, if impalpably fine, clay. These fragmental rocks cannot be classified chemically, since the same handful of gravel, for instance, may contain pebbles of many different kinds of rocks, and thus be of almost any and very variable composition. Such chemical distinctions as can be established are only partial, and the classification, like the origin, must be mechanical. Accordingly, as just shown, we recognize three principal groups based upon the size of the fragments; viz.:—

    (1) Conglomerate group.
    (2) Arenaceous group.
    (3) Argillaceous group.

This mode of division is possible and natural, simply because, as we observed in an early experiment, materials arranged by the mechanical action of water are always assorted according to size. When first deposited, the gravel, sand, and clay are, of course, perfectly loose and unconsolidated; but in the course of time they may, under the influence of pressure, heat, and chemical action, attain almost any degree of consolidation, becoming *conglomerate, sandstone,* and *slate,* respectively. The pressure may be vertical where it is due to the weight of newer deposits, or horizontal where it results from the cooling and shrinking of the earth's interior. The heat may result from mechanical movements, or contact with eruptive rocks; or it may be due simply to the burial of the sediments, which, it will be seen, must virtually bring them nearer the great source of heat in the earth's interior, on the same principle that the temperature of a man's coat, on a cold day, is raised by putting on an overcoat. The effect of the heat, ordinarily, is simply drying, coöperating with the pressure to expel the water from the sediments; but, if the temperature is high, it may bake or vitrify them, just as in brick-making. Sediments are consolidated by chemical action when mineral substances, especially calcium carbonate, the iron oxides, and silica are deposited between the particles by infiltrating waters, cementing the particles together. This principle is easily demonstrated experimentally by taking some loose sand and wetting it repeatedly with a saturated solution of some soluble mineral, like salt or alum, allowing the water to evaporate each time before making a fresh application. The interstices between the grains are gradually filled up, and the sand soon becomes a firm rock. But the student should clearly understand that, in geology, gravel, sand, and clay are just as truly *rocks* before their consolidation as after. It is plain then that in each of the principal groups of fragmental rocks we must recognize an unconsolidated division and a consolidated division.

(1) *Conglomerate group.*—The rocks belonging in this group we know before consolidation as *gravel*, and after consolidation as *conglomerate.*

Gravel.—The pebbles, as we have already seen, are usually, though not always, well rounded or water-worn; and they may be of any size from coarse grains of sand to boulders. As a rule, however, the larger pebbles, especially, are of approximately uniform size in the same bed or layer of gravel, with, of course, sufficient fine material to fill the interstices. Although the same limited mass of gravel may show the widest possible range in chemical and mineralogical composition, yet hard rocks are evidently more likely than soft rocks to form pebbles; and hence quartz and quartz-bearing rocks usually predominate in gravels. Specimen 28.

Conglomerate.—Consolidated gravel. Children should be led to an appreciation of this point by a careful comparison of the forms of the pebbles in the gravel and conglomerate. The conglomerate seems to contain a larger proportion of fine material than ordinary gravel. But this is because the gravel is usually, as with our specimen, taken from the *surface* of the beach, where, of course, the pebbles are clean and separate; but if it had remained there to be covered by a subsequently deposited layer, enough fine stuff would have been sifted into the holes to fill them. And in the finished gravel, just as in the conglomerate, the pebbles are usually closely packed, with just sufficient sand and clay, or *paste*, as the material in which the pebbles are imbedded is called, to fill the interstices. The paste is usually similar in composition to the pebbles, with this difference: hard materials predominate in the pebbles and soft in the paste.

Stratified rocks generally show the stratification in parallel lines or bands differing in color, composition, etc.; but nothing like this can be detected in our specimens of conglomerate; and the question might be asked, How do we know that this is a stratified rock? In answer, it can be said that our hand-specimens appear unstratified simply because the rock is so coarse; but when we look at large masses, and especially when we see it in place in the quarry, that parallel arrangement of the material which we call stratification is usually very evident; and we often see precisely the same thing in gravel banks. It is, however, wholly unnecessary that we should *see* the stratification in order to know certainly that this is a stratified or aqueous rock, because the forms of the pebbles show very plainly that they have been fashioned and deposited by moving water; and we have in the smallest specimen proof positive that our conglomerate is a consolidated sea-beach.

Conglomerate shows the same variations in composition and texture as gravel; it may be composed of almost any kind of material in pebbles of almost any size. We recognize two principal varieties of conglomerate based on the forms of the pebbles; if, as is usual, these are well rounded and water-worn, the rock is true *pudding-stone* (specimen 29); but, if they are angular, or show but little wear, it is called *breccia.*

(2) *Arenaceous Group.*—The conglomerate group passes insensibly into the arenaceous group; for, from the coarsest gravel to the finest sand, the gradation is unbroken, and every sandstone is merely a conglomerate on a small scale.

Sand.—Like gravel, sand may be of almost any composition, but as a rule it is quartzose; quartz, on account of its hardness and the absence of cleavage, being better adapted than any other common mineral to form sand. Where the composition of a sand is not specified, a quartzose sand is always understood. By examining a typical sand with a lens, and noting the glassy appearance of the grains, and then testing their hardness on a piece of glass, which they will scratch as easily as quartz, the pupil is readily convinced that each grain is simply an angular fragment of quartz. Specimen 30.

Sandstone.—Consolidated sand. In proving this, children will notice first the granular or sandy appearance of the sandstone; and then, with the lens, that the grains in the sandstone have the same forms as the sand-grains. The stratification cannot be seen very distinctly in our hand-specimens, but in larger masses it is usually very plain, as may be observed in the blocks used for building, and still better in the quarries. However, even if the stratification were not visible to the eye, we could have no doubt that sandstone is a mechanically formed stratified rock; because the form of the grains, just as in the conglomerate, tells us that. Many sandstones, too, contain the fossil remains of plants and animals, and these are always regarded as affording positive proof that the rocks containing them belong to the aqueous or stratified series.

There are many varieties of sandstone depending upon differences in composition, texture, etc., but we have not space to notice them in detail. In sandstone, just as in sand, quartz is the predominant constituent, although we sometimes find varieties composed largely or entirely of feldspar, mica, calcite, or other minerals. Specimen 31 is an example of the architectural variety known as freestone, which is merely a fine-grained, light-colored, uniform sandstone, not very hard, and breaking with about equal freedom in all directions. The consolidation of sandstones is due chiefly to chemical action. The cementing materials are commonly either: *ferruginous* (iron oxides), giving red or brown sandstones; *calcareous*, forming soft sandstones, which effervesce with acid if the cement is abundant; or *siliceous*, making very strong, light-colored sandstones. Ferruginous sandstones are the most valuable for architectural purposes; for, while not excessively hard, they have a very durable cement. Siliceous sandstones are too hard; and the calcareous varieties crumble when exposed to the weather because the cement is soluble in water containing carbon dioxide, as all rain-water does. Specimen 32 is a good example of a ferruginous sandstone, and it is coarse enough so that we can see that each grain of quartz is coated with the red oxide of iron. The mica scales visible here and there in this specimen are interesting as showing that the grains are not necessarily all quartz; and it is important to observe that the mica was not made in the sandstone, but, like the quartz, has come from some older rock.

Quartzite.—This rock is simply an unusually hard sandstone. Now the hardness of any rock depends upon two things: (1) the hardness of the individual grains or particles; and (2) the firmness with which they are united one to another. Therefore, the hardest sandstones must be those in which grains of quartz are combined with an abundant siliceous cement; and that is precisely what we have in a typical quartzite, such as specimen 33. Quartzite is distinguished, in the hand-specimen, from ordinary quartz by its granular texture (compare specimens 15 and 33); and of course in large masses the stratification is an important distinguishing feature.

3. *Argillaceous group.*—Just as the conglomerate group shades off gradually into the arenaceous group, so we find it difficult to draw any sharp line of division between the arenaceous group and the argillaceous, but we pass from the largest pebble to the most minute clay-particle by an insensible gradation. For the sake of convenience, however, we draw the line at the limit of visibility, and say that in the true clay and slate the individual particles are invisible to the naked eye; in

other words, these rocks have a perfectly compact texture, while the two preceding groups are characterized by a granular texture. Although clay, like sand and gravel, may be of almost any composition, yet it usually consists chiefly, often entirely, of the mineral kaolin. The reason for this is easily found. Quartz resists both mechanical and chemical forces, and is rarely reduced to an impalpable fineness; but all the other common minerals, such as feldspar, hornblende, mica, and calcite, on account of their cleavage and inferior hardness, are easily pulverized; but it is practically impossible that this should happen without their being broken up chemically at the same time. Decomposition follows disintegration; and, while calcite is completely dissolved and carried away, the other minerals are reduced, as we have seen, to impalpable hydrous silicates of aluminum, *i.e.*, to kaolin. Hence, we find that the fragmental rocks are composed principally of two minerals, quartz and kaolin, — the former predominating in the conglomerate and arenaceous groups, and the latter in the argillaceous group.

Clay.—That kaolin is the basis of common clay is proved by the argillaceous odor, which is so characteristic of moist clay. Pure kaolin clay is white and impalpable, like China clay; but pure clays are the exception. They often become coarse and gritty by admixture with sand, forming *loam*; and they also usually contain more or less carbonaceous matter, which makes black clays; or more or less *ferrous* oxide, which makes blue clays; or more or less *ferric* oxide, which makes red, brown, and yellow clays. By mixing these coloring materials in various proportions, almost any tint may be explained. Clays are sometimes calcareous, from the presence of shells and shell-fragments or of pulverized limestone. These usually effervesce with acid, and are commonly known as *marl*. It is the calcareous material in a pulverulent and easily soluble condition that makes the marls valuable as soils.

Slate.—Consolidated clay. The compact texture and argillaceous odor are usually sufficient to prove this. To get the odor we need simply to breathe upon the specimen, and then smell of it. We find all degrees of induration in clay. It sometimes, as every one knows, becomes very hard by simple drying; but this is not slate, and no amount of mere drying will change clay into slate; for, when moistened with water, the dried clay is easily brought back to the plastic state. To make a good slate, the induration must be the result of pressure, aided probably to some extent by heat. True slate, then, is a permanently indurated clay which will not soak up and become soft when wet.

Slate is usually easily scratched with a knife, and it is distinguished from limestone by its non-effervescence with acid. As we should expect, it shows precisely the same varieties in color and composition as clay. A good assortment of colors is afforded by the roofing-slates. Specimen 34 is a typical slate, for it not only has a compact texture and argillaceous odor, but it is very distinctly stratified. The stratification is marked by alternating bands of slightly different colors, and is much finer and more regular than we usually observe in sandstone, and of course entirely unlike the stratification of conglomerate. These differences are characteristic. Some slates, however, are so homogeneous that the stratification is scarcely visible in small pieces. Thus the roofing-slates (specimen 35) rarely show the stratification; for it is an important fact that the thin layers into which this variety splits are entirely independent of the stratification. This is the structure known as slaty cleavage; it is not due to the stratification,. but is developed in the slate subsequently to its deposition, by pressure. Some roofing-slates, known as ribbon-slates, show bands of color across the flat surfaces. These bands are the true bedding, and indicate the absolute want of conformity between this structure and the cleavage. Few rocks are richer in fossils than slate, and these prove that it is a stratified rock. Slate which splits easily into thin layers *parallel with the bedding* is known as *shale.*

Porcelainite.—This is clay or slate which has been baked or partially vitrified by heat so as to have the hardness and texture of porcelain.

2. Chemically and Organically formed Rocks.—We have already learned that from a geological point of view the differences between chemical and organic deposition are not great, the process being essentially chemical in each case; and since the limestones and some other important rocks are deposited in both ways, it is evidently not only unnatural, but frequently impossible, to separate the chemically from the organically formed rocks. Unlike the fragmental rocks, the rocks of this division not only admit, but require, a chemical classification. This is natural because they are of chemical origin; and it is practicable because, with few exceptions, only one class of minerals is deposited at the same time in the same place,—a very convenient and important fact. Therefore our arrangement will be mineralogical, thus:—

(1) Coal Group.
(2) Iron-ore Group.
(3) Siliceous Group.
(4) Calcareous Group.
(5) Metamorphic Group (Silicates).

Most of the silicate rocks are mixed, *i.e.,* are each composed of several minerals; but some silicate rocks and all the rocks of the other divisions are simple, each species consisting of a single mineral only.

(1) *Coal Group.*—These are entirely of organic origin, and include two allied series, which are always merely the more or less extensively transformed tissues of plants or animals; viz.:—

Coals and **Bitumens.**—At the first lesson we examined a sample of peat (specimen 8), and considered the general conditions of its formation, peat being in every instance simply partially decayed marsh vegetation. It was also stated that, as during the lapse of time the transformation becomes more complete, the peat is changed in succession to *lignite, bituminous coal, anthracite,* and *graphite.* The coals, indeed, make a very beautiful and perfect series, whether we consider the composition—there being a gradual, progressive change from the composition of ordinary woody fibre in the newest peat to the pure carbon in graphite,—or the degree of consolidation and mineralization—since there is a gradual passage from the light, porous peat, showing distinctly the vegetable forms, to the heavy crystalline graphite, bearing no trace of its vegetable origin. This relation is easily appreciated by a child, if a proper series of specimens is presented. The coals also make a chronological series, graphite and anthracite occurring only in the older formations, and lignite and peat in the newer, while bituminous coal is found in formations of intermediate age.

Bituminous coal is the typical, the representative coal; and from a good specimen of this variety we may learn two important facts:—

(1) That true coals, no less than peat, are of vegetable origin. To see this we must look at the flat or charcoal surfaces of the coal. These soil the fingers like charcoal, and usually show the vegetable forms distinctly.

(2) That coals are stratified rocks. These dirty charcoal surfaces always coincide with the stratification, being merely the successive layers of vegetation deposited and pressed together to build up the coal; and when we look at the edge of the specimen the stratification shows plainly enough.

The bitumens form a similar though less perfect series, beginning with the organic tissues, and ending, in the opinion of some of the best chemists and mineralogists, with diamond. In fact the coals and bitumens form two distinct but parallel series. The coals are exclusively of vegetable origin, while the bitumens are largely of animal origin. The organic tissues in which the two series originate are chemically similar,—the animal tissues, which produce the lighter forms of bitumen, however, containing more hydrogen and less carbon and oxygen than vegetable tissues; while the final terms, as just shown, are probably chemically identical, being pure carbon,—graphite for the coals and diamond for the bitumens; so that the entire process of change in each series is essentially carbonization, a gradual elimination of the gaseous elements, oxygen and hydrogen, until pure solid carbon alone remains.

The principal differences between the coals and bitumens are the following:—

Coals are rich in carbon, with some oxygen and little hydrogen.

Bitumens are rich in hydrogen, with some carbon and little or no oxygen.

Coals are entirely insoluble.

Bitumens are soluble in ether, benzole, turpentine, etc., and the solid forms are soluble in the more fluid, naphtha-like varieties.

Coals are never liquid, and cannot be melted or, with trifling exceptions, even softened by heat.

Many bitumens are naturally liquid, and all become so on the application of heat.

The coals partake of the characteristics of their chief constituent element, carbon, the most thoroughly solid substance known; while the bitumens similarly show the influence of hydrogen, the most perfectly fluid substance known.

The two bitumens of the greatest geological importance are asphaltum or mineral pitch and petroleum; but these substances are too familiar to require any farther description here.

(2) *Iron-ore Group.*—These interesting and important stratified rocks include the three principal oxides of iron,—limonite, hematite, and magnetite,—as well as the carbonate of iron, siderite; and the rocks have essentially the same characteristics as the minerals. In economical importance they are second only to the coals; and the history of their formation through the agency of organic matter is one of the most interesting chapters in chemical geology (see page 26). The three oxides are easily distinguished from each other by the colors of their powders or streaks, and the magnetism of magnetite, and from all other common rocks by their high specific gravity. Magnetite is the richest in iron, and limonite the poorest. As regards the degree of crystallization and order of occurrence in the formations, they form a series parallel with the coal series, thus:—

Limonite, never crystalline, and found in recent formations.
Hematite, often crystalline, and found in older formations.
Magnetite, always crystalline, and found in oldest formations.

Siderite effervesces with strong acid; and this separates it from all other rocks, except limestone and dolomite; and from these it is distinguished by its high specific gravity. As a mineral, siderite is often light colored; but as a rock it is always dark, and usually black, from admixture chiefly of carbonaceous matter. In studying dynamical geology, we have learned (page 28) the reason for the intimate association of siderite with beds of coal, and this accounts equally for the carbon contained in the rock itself. The connection of this rock with the coal-formations adds much to its value as an ore of iron.

Finally, the iron-ores, at least where of much economical importance, are truly stratified. This can often be seen in hand-specimens; and is well shown by their relations to other rocks, in quarries and mines; and in many cases, for limonite and hematite, by the fossils which they contain.

(3) *Siliceous Group.*—These rocks are composed of pure silica in the forms of quartz and opal. When first deposited, whether organically, like tripolite, or chemically, like siliceous tufa, the siliceous rocks are soft and light, and the silica is in the form of opal. Subsequently it changes to quartz, and the rocks assume the much harder and denser forms of chert and novaculite, respectively.

Tripolite or **Diatomaceous Earth.**—This interesting rock is soft, light, and looks like clay; but it is lighter, and the argillaceous odor is faint or wanting. It does not effervesce with acid. Hence, it is neither clay nor chalk. Notwithstanding its softness, it is really composed of a hard substance, viz., silica, in the form known as opal. By rubbing off a little of the dust, and examining it under the microscope, we easily prove that the silica is mainly or entirely of organic origin; for the dust is seen to be simply a mass of more or less fragmentary organic remains, occurring in great variety, and of wonderful beauty and minuteness. There are few rocks so unpromising on the exterior, and yet so beautiful within. We have already learned that these organic bodies are principally Diatom cases, Radiolaria shells, and Sponge spicules. We can form some idea of their minuteness from Ehrenberg's estimate that a single cubic inch of pure tripolite contained no less than 41,000,000,000 organisms.

The lightness of tripolite (sp. gr., 1-1.5) is due to the facts that opal is a light mineral (sp. gr., 1.9-2.2), and that many of the shells are hollow. Tripolite is a good example of a soft rock composed of a hard mineral; and it owes its value as a polishing material to the fact that it consists of a hard mineral in an exceedingly fine state of division. Tripolite, when pure, is snow-white; but it is rarely pure, being commonly either argillaceous or calcareous. This rock is now forming in thousands of places, in both fresh water and the ocean.

Flint and **Chert**.—During the course of geological time, beds of tripolite are gradually consolidated, chiefly by percolating waters, which are constantly dissolving and re-depositing the silica; and, finally, in the place of a soft, earthy rock, we get a hard, flinty one, which we call *flint* if it occurs in the newer, or *chert* if it occurs in the older, geological formations. Besides forming beds of nearly pure silica, which we call tripolite, the microscopic siliceous organisms are diffused more or less abundantly through other rocks, especially chalk and limestone. In such cases the consolidation of the silica implies its segregation also; *i.e.*, the silica dissolved by percolating water is deposited only about certain points in the rock, building up rounded concretions or nodules. Thus, a siliceous limestone becomes, by the segregation of the silica, a pure limestone containing nodules of chert, which are usually arranged in lines parallel with the stratification. Specimen 16 is a fragment of a flint-nodule from the chalk-formation of England.

Siliceous Tufa.—Hot water, and especially hot alkaline water, circulating through the earth's crust, is always charged with silica dissolved out of the rocks; and when such water issues on the surface in a hot spring or geyser, it is cooled by contact with the air, its solvent power is diminished thereby, and a large part of the silica is deposited around the outlet as a snowy-white porous material called *siliceous tufa*. Silica deposited in this way is, like organic silica, always in the form of opal. Siliceous tufa is distinguished from clay, slate, chalk, and limestone by the same tests as tripolite, and from tripolite itself by the absence of microscopic organisms.

Novaculite.—Through the action of percolating water and pressure, siliceous tufa, like tripolite, becomes harder and denser and is thus changed to *novaculite*, which holds the same relation to chemically deposited silica that chert and flint do to organically deposited silica. The white novaculite obtained at the Hot Springs of Arkansas, and commonly known as Arkansas stone, is a typical example of this rock. The rock which, on account of the use to which it is put, is known as buhr-stone, is also an excellent example of chemically deposited silica. It is usually somewhat porous and fossiliferous.

(4) *Calcareous Group*.—These are the lime-rocks, including the carbonate of lime, in limestone and dolomite, the sulphate of lime, in gypsum, and the phosphate of lime, in phosphate rock. These rocks are even more closely connected in origin than in composition; and it is for this reason that rock-salt, which of course

contains no lime, is also included in this group. Limestones are formed abundantly in the open sea, through the accumulation of shells and corals; but when portions of the sea become detached from the main body and gradually dry up, like the Dead Sea and Great Salt Lake, dolomite, gypsum, and rock-salt are deposited in succession as chemical precipitates. Phosphate rock may be regarded as a variety of limestone, resulting from the accumulation of the skeletons and excrement of the higher animals.

Limestone.—This is the lithologic or rock form of carbonate of lime or calcite, and one of the most important, interesting, and useful of all rocks. Although so simple in composition,—calcite being the only essential constituent,—limestone embraces many distinct varieties, and is really equivalent to a whole family of rocks. A highly fossiliferous limestone, such as specimen 38, is, perhaps, the best variety with which to begin the study of the species. The softness of the fossil shells of which the rock is so largely composed, and the fact that they effervesce readily with dilute acid, proves that they are still carbonate of lime; and by applying the acid more carefully, it can be seen that the compact matrix of the rock also effervesces, consisting of shells more finely broken or comminuted and mixed with more or less clay and other impurity, almost the entire rock being of organic origin.

On the coast of Florida, and in many other places, we find beautiful examples of shell-limestone now in process of formation. These are at first very open and porous, because the interstices between the nearly entire shells are not yet filled up with smaller fragments and sand. But when that is done, we shall have a rock similar to the old fossiliferous limestone. Specimen 37.

The shells and fragments, and the grains of calcareous sand, are, as a rule, quickly cemented together by the deposition of carbonate of lime between them; so that limestone is nowhere observed occurring abundantly in an unconsolidated form.

Limestone, as a rule, is not distinctly stratified in hand-specimens, but of course that it is a true sedimentary rock is abundantly proved by the fossils; and it goes almost without saying that limestone, being necessarily mainly composed of organic remains, must be to a greater extent than any other rock the great storehouse of fossils; and in no other rock are the fossils so well preserved and perfect as in limestone.

Nevertheless, there are extensive formations of limestone containing no discernible traces of fossils. And some of these non-fossiliferous limestones, too, are of very recent formation. Some of the modern coral-reefs, for example, are composed of limestone which was formed only yesterday, as it were, and which, from its mode of formation, must consist entirely of corals; and yet it shows no trace of its organic origin, but is perfectly compact, or, possibly, crystalline. This frequent obliteration of the organic remains, as well as the perfect consolidation of the rock, is attributed to its solubility. The calcium carbonate is gradually dissolved by the water, and then deposited in the interstices in other parts of the rock.

Specimen 39 is that variety of limestone known as *chalk*. It is soft and earthy, resembling both clay and tripolite, but differing from the former in lacking the distinct argillaceous odor, and from both by its lively effervescence with acids. It appears to be entirely destitute of organic remains, but this is a defect of our vision and not of the rock; for, like the tripolite, it often appears under the microscope to be little else than a mass of shells. Tripolite is a deposit built up of the siliceous shells of Diatoms and Radiolaria, while chalk is chiefly composed of the similar but calcareous shells of Foraminifera. Our specimen is from the Cretaceous formation of England; but we have good reason to believe that chalk is *now forming* on a very extensive scale. There are millions of square miles in the deeper parts of the ocean where the dredge brings up little else but a perfectly impalpable, gray, calcareous slime or ooze. When examined microscopically, this is seen to be composed chiefly of Foraminifera shells, and among these the genus Globigerina predominates; so that the deposit is frequently called Globigerina ooze. Now this gray, calcareous ooze, when dried and compacted by pressure, makes a soft, *white* rock which can scarcely be distinguished from chalk; in fact, it is a modern chalk. And there seems no good reason to doubt that the deposition of chalk has gone on continuously since Cretaceous time—for several millions of years at least.

Specimen 40 is also a white rock, easily scratched with the knife, and effervescing freely with acid, and therefore a variety of limestone. But its texture is very different from the other varieties we have studied. It has a sparkling surface, which we explain by saying that the rock is crystalline. It is, in fact, a mass of minute crystals of calcite. The crystalline limestones have not always been crystalline, but it is safe to assume that they were originally entirely uncrystalline, and in many cases rich in fossils; but the fossils have been mainly obliterated by the crystallization.

Crystallization generally in rocks is an indication of great age, so that we usually say crystalline rocks must be older than uncrystalline rocks of the same composition; and this is mainly true with the limestones. When the crystallization is rather fine, as in our specimen, resembling granulated sugar, we have what is commonly called saccharoidal limestone. This is the typical marble. Marble is not a scientific name, and the term is usually applied to any calcareous rock which will take a polish, and sometimes even to rocks which are not calcareous at all.

In the section on dynamical geology, we learned that the carbonate of calcium or calcite is deposited from the sea-water, and limestones formed, in two ways: first, in a purely chemical way, where the water becomes saturated with calcite; and, second, organically, where the calcium carbonate is taken from the water by marine organisms to form their shells and skeletons, and the gradual accumulation of these on the ocean-floor builds up a limestone. As before stated, the difference between these two methods of deposition is not so great as it often seems, because we know that the animals never make the carbonate of calcium which they secrete, but it comes into the sea ready made with the drainage from the land.

The limestones forming at the present time are almost wholly organic; but the rock known as *calcareous tufa* is an exception. This is formed under the same general conditions as siliceous tufa, but much more abundantly, and in cold water as well as warm; because calcite is far more soluble (especially in water containing carbon dioxide) than opal or quartz. It is deposited, not only around the mouths of springs, but also along the beds of the streams which they form, enveloping stones, roots, grasses, etc., and building up usually a loose, spongy mass having a very characteristic turfaceous texture.

The principal accessory minerals occurring in limestone are: (1) *kaolin*, forming argillaceous or slaty limestone, which may be recognized by the argillaceous odor and dark color; (2) *quartz*, forming siliceous or cherty limestone, known by its hardness or by the nodules of flint or chert; (3) *dolomite*, forming dolomitic or magnesian limestone, which effervesces less freely with acid; and (4) *serpentine*, forming serpentinic limestone, which is sharply distinguished by the green grains of serpentine mingled with the white calcite. A concretionary texture is common with limestone. If the concretions are small, like mustard-seed, we call the rock *oölite*; if larger, like peas, *pisolite*.

Dolomite.—If for calcite, which is the sole essential constituent of all limestone, we substitute the allied mineral dolomite, we have the rock dolomite. As might be inferred from its composition, dolomite is very closely related to limestone, although there are some important differences. Physically, the two rocks differ about as the two minerals do. Dolomite is harder than limestone, and being also less soluble, it resists the action of the weather more. Dolomite, if pure, effervesces feebly, or not at all, with cold dilute acid. Here, however, we have to recognize the fact that dolomite is rarely pure; but there exists, in consequence of the admixture of calcite, a perfectly gradual passage from pure dolomite to pure limestone, and parallel with this every degree of vigor in the reaction with acid. Hence, it is entirely an arbitrary matter as to where we shall draw the line between dolomitic limestone and calcareous dolomite. Dolomite is a very much less abundant rock than limestone, and, unlike limestone, it rarely contains many fossils, and is never of organic origin; *i.e.*, there are no organisms which secrete the mineral dolomite to form their hard parts or skeletons. Like gypsum and rock-salt, dolomite is probably never deposited in the open ocean, but only in closed basins. Like limestone, it occurs with both the compact and the crystalline textures.

Gypsum.—When pure, this rock (specimen 36) is identical with the mineral gypsum (specimen 17), except that it is rarely crystalline. It is usually, however, not only perfectly compact, but more or less dark-colored from the admixture of clay and other impurities. Its most notable characteristics are its softness, the absence of the argillaceous odor, except where it contains much clayey impurity, and its non-effervescence with acids. The first two usually serve to distinguish it from slate, while the acid test separates it readily from limestone and all other carbonate rocks. The deposition of gypsum is purely chemical, and it occurs under about the same physical conditions as the deposition of salt; *i.e.*, in drying-up portions of the sea. Hence we usually find gypsum associated with beds of rock-salt; and, since drying-up seas are few in number, and small compared with the whole extent of the ocean, we can easily understand why neither rock-salt nor gypsum are abundant rocks, except in a few localities.

Rock-Salt.—This interesting and useful rock, as we have already learned, is deposited in a purely chemical way, and only in drying-up portions of the sea, like the Dead Sea, Great Salt Lake, etc. In some parts of Europe there are beds of solid rock-salt over a hundred feet thick.

Phosphate Rock.—Although not specially abundant or attractive, this rock is of great economic interest and importance on account of its extensive use as a fertilizer. Under the general head of phosphate rock are included: (1) the typical guano, which is the consolidated excrement of certain marine birds inhabiting in great numbers small coral islands in the dry or rainless regions of the tropics; (2) the underlying coral rock, which is often changed to phosphate rock through the percolation of the rain-water falling on the guano; (3) accumulations of the bones and coprolites of the higher animals; (4) phosphatic limestones from which the carbonate of lime has been largely dissolved away, leaving the more insoluble phosphate of lime.

(5) *Metamorphic Group* (stratified silicates).—All the chemically and organically formed rocks which we have studied up to this point are simple, *i.e.*, they consist each of only one essential mineral; but most of the rocks in this great group of silicates are mixed, or consist each of several essential minerals. Quartz is the only important constituent of these rocks which is not, strictly speaking, a silicate, but in a certain sense it is also not an exception, since it may always be regarded as an excess of acid in the rock.

This group of stratified rocks composed of silicate minerals is of exceptional importance, first, on account of the large number of species which it includes, and, second, on account of the vast abundance of some of the species. These are, above all others, the rocks of which the earth's crust is composed. With unimportant exceptions, all the rocks of this group are crystalline; and they constitute the principal part of what is generally included under the term *metamorphic rocks*—a general name for all stratified rocks which have been so acted upon by heat, pressure, or chemical forces as to make them crystalline. Although the crystalline limestone, dolomite, iron-ores, etc., show us that metamorphic rocks are not wanting in the other groups.

As already explained, the metamorphic or crystalline stratified rocks are usually older than the corresponding uncrystalline rocks; but a point of greater importance here is this: the development in the silicate rocks of crystalline characters has usually made it impossible to determine the method of their deposition, whether mechanical or chemical. In a few cases, as with the rock greensand, we know that the deposition is chemical; while it is equally certain that such common silicate rocks as gneiss, mica schist, and many others, often result from the crystallization of ordinary mechanical sediments, like sandstone and conglomerate. We classify all these rocks as of chemical origin, however, without considering the mode of their deposition, because the subsequent crystallization is itself essentially a chemical process; and that justifies us in saying that these rocks are made what they now are chiefly by the action of chemical forces. Whatever they were originally, they have become, through their crystallization, rocks having a definite mineral composition which can be classified chemically.

Some of the details of the classification of this group, as shown in the table, require explanation. In studying the silicate minerals it was stated to be important to recognize two classes—the *acidic* and the *basic*—the dividing line falling in the neighborhood of 60 per cent. of silica. This division is important simply because Nature has in a great degree kept the acidic and basic minerals separate in the rocks; and few things in lithology are more important than the distinction of the silicate rocks in which acidic minerals predominate from those in which basic minerals predominate. The amount of silica which any rock of this group contains is shown at a glance by the chart. The vertical broken lines, with the figures at the top, indicate the proportion of silica, which increases from 30 per cent. on the right to 85 per cent. on the left; so that the percentage of silica which a rock contains determines its position, the acidic species being on the left, and the basic on the right. As most of these rocks are composed of two or more minerals mixed in very various proportions, there is usually a wide range in the percentage of silica which the same species may contain; and this is expressed in each case by the length of the dotted line under the name of the rock. Thus, in syenite, the silica ranges from 55 per cent. to 65 per cent. The horizontal line in the chart separates the gneisses, containing feldspar as an essential constituent, from the schists, in which feldspar is wanting, except as an accessory constituent. We will take up the gneisses first.

Gneiss.—This is the most important of all rocks. It forms not far from one-half of New England, and a very large proportion of the earth's crust. The name (pronounced same as *nice*) is known to have originated among the Saxon miners, but its precise derivation is lost in obscurity. To find out what this very important rock is, we will consult specimen 41. The first glance shows us that it is not, like the rocks we have just been studying, composed of a single mineral, but of several minerals, the most conspicuous of which is the pink feldspar—orthoclase. This we recognize as a feldspar: (1) by its hardness, which is a little less than that of quartz, and distinguishes it from calcite, a mineral having the general appearance of feldspar; (2) by its color, which separates it from hornblende and augite; and (3) by its cleavage, which distinguishes it easily from quartz. Finally, we know it is orthoclase, and not plagioclase, by its general aspect, and by its association with an abundance of quartz, which is the next most important constituent of the rock. The quartz is less abundant than the orthoclase, and more easily overlooked, yet anyone familiar with the mineral will not fail to recognize it. It forms small, irregular, glassy grains, entirely devoid of cleavage, and scratching glass easily. On weathered surfaces of the rock the orthoclase becomes soft and chalky, while the quartz remains clear and hard, and then the two minerals are very easily distinguished. Besides these, there are numerous black, thin, glistening scales, which we can easily prove to be elastic, and recognize as mica.

In most books on the subject, these three minerals—orthoclase, quartz, and mica—are set down as the normal or essential constituents of gneiss. But it is now recognized by the best lithologists that we may have true gneiss without any mica; or we may have hornblende in the place of mica. Quartz and orthoclase are the only essential constituents of gneiss; and when these alone are present, we have the variety known as binary gneiss. The addition to these essential constituents of mica, gives micaceous gneiss; and of hornblende, hornblendic gneiss. Of these three principal varieties, the micaceous gneiss is by far the most common and important. The mica may be either the white species, muscovite, or the black species, biotite; but it is usually the former.

Orthoclase is the predominant constituent in all typical gneiss, usually forming at least one-half of the rock. The orthoclase may, however, be replaced to a greater or less extent by albite, or even by oligoclase. But we frequently see the term *gneiss* carelessly, or ignorantly, applied to rocks which are destitute of feldspar, though having the general aspect of gneiss.

Augite rarely occurs in gneiss; and hence, when we observe a gneiss containing a black mineral which we know is either augite or hornblende, it is pretty safe to call it the latter.

Mica and hornblende, although the principal, are not the only, accessory minerals in gneiss; but the following species are also of common occurrence: garnet, cyanite, tourmaline, fibrolite, epidote, and chlorite. Gneisses, as the table indicates, exhibit a wide range in the proportion of silica which they contain, varying from 60 to 85 per cent.; and there is a concomitant variation in specific gravity, from about 2.5 in the most acidic to 2.8 in the most basic varieties.

That gneiss is a true, stratified rock is very clearly shown in specimen 41; but, unfortunately, the stratification is not always so evident as in this case. The mica-scales, it will be observed, lie parallel with the stratification, and assist very materially to make it visible; and gneisses containing little or no mica, as well as some that are rich in mica, frequently appear almost or quite unstratified. These obscurely stratified varieties are commonly known as granitoid gneiss, having the texture and general aspect of granite. The sedimentary origin of gneiss is also clearly proved by its interstratification with undoubted sedimentary rocks, such as limestone, iron-ores, graphite, quartzite, etc.

Syenite.—This is a much abused term, but, as now employed by the best lithologists, it is the name of a rock having a single essential constituent, viz., orthoclase. Syenite in its simplest variety contains nothing but orthoclase; but in addition we usually have either hornblende, forming hornblendic syenite, or mica, forming micaceous syenite.

Syenite, it will be observed, is equivalent to gneiss with the quartz removed; but, while gneiss is the most abundant of all rocks, syenite is a comparatively rare rock; and this is simply another way of saying that nearly all orthoclase is associated with quartz. By admixture of quartz we get a perfectly gradual passage from syenite to gneiss. The orthoclase in syenite is more frequently replaced by plagioclase than it is in gneiss. In syenite, too, hornblende is much more abundant than mica; although just the opposite is true in gneiss. And, again, in gneiss the mica is principally muscovite; but in syenite it is almost exclusively biotite. Augite is a common accessory in the more basic syenite; but garnet, tourmaline, and the other accessory minerals, occurring so frequently in gneiss, are almost unknown in syenite. The specific gravity of syenite varies from 2.7 to 2.9.

Diorite.—This is a more important rock than syenite; but it is of analogous, though more basic, composition, containing a single essential constituent, viz., plagioclase. Any of the triclinic feldspars may occur in this rock, but oligoclase is most common. Like syenite, diorite usually contains hornblende, often in large proportion, forming hornblendic diorite, which sometimes passes into rocks composed entirely of hornblende. It also, but less frequently, contains mica, forming micaceous diorite. The mica is usually biotite, rarely muscovite. Mica and hornblende also often occur together in diorite, and the same is true of syenite and gneiss. Quartz is of common occurrence in the more acidic varieties of diorite, and augite in the more basic.

This is a good example of a basic rock, for all its normal constituents are basic; but the percentage of silica varies from 45 in those varieties richest in labradorite and augite to 60 or more in those containing more or less quartz and orthoclase. There is a corresponding change of color from dark to light, and of specific gravity from 2.7 to 3.1.

Diorite is not rich in accessory minerals; besides those already mentioned, the most important are chlorite, epidote, pyrite, and magnetite.

Few rocks are more clearly stratified than diorite, whether we consider the hand-specimen, or its relations to other formations. It is an abundant rock in New England.

Norite.—Like diorite, this is essentially a plagioclase rock; but there are, nevertheless, important differences. The plagioclase in diorite is mainly the more acidic species, like oligoclase; while in norite the more basic species, such as labradorite and anorthite, predominate. Hornblende, which we have observed to be an important and rather constant constituent of diorite and syenite, is much less abundant in norite; but its place is taken by augite and the allied minerals, hypersthene and enstatite. Black mica is common in norite; but white mica, orthoclase, and quartz rarely occur.

Norite is the most basic of all the feldspathic rocks, as gneiss is the most acidic; while syenite and diorite stand as connecting links, forming a gradual passage between the two extremes. Thus, in passing from gneiss to norite, we have observed a gradual diminution of the quartz, a gradual change in feldspar from orthoclase to the most basic plagioclase; at first a gradual increase in hornblende, and then a gradual change from hornblende to augite; and, finally, a gradual substitution of black mica for white. The amount of silica has decreased over 40 per cent.; and the specific gravity has increased from 2.5 in the lightest gneiss to at least 3.2 in the heaviest norite. We have also passed from light colored rocks to dark; and from those resisting atmospheric action to those easily decomposed.

The most characteristic accessory constituents of norite, besides those already mentioned, are magnetite and chrysolite; though garnet, serpentine, and pyrite often occur. In texture, this rock varies from compact to very coarsely crystalline. The specimen of labradorite (No. 23), from the norite of Labrador, affords some idea of the coarseness of the crystallization in much of this rock. It is not a common rock, except in certain regions, the best known of which in eastern North America are the coast of Labrador, various points in Canada north of the St. Lawrence, and the eastern border of the Adirondack Mountains. In hand-specimens, norite rarely appears stratified; but in the solid ledges the stratification is often as distinct as could be desired.

Many lithologists call the rocks here designated norite *gabbro*, and class them all in the eruptive division as essentially a coarse variety of diabase. In a similar manner, diorite and syenite are denied a place in the sedimentary series. But the stratified plagioclase rocks seem to have as strong a claim to recognition as gneiss.

We turn now to the important and interesting division of the non-feldspathic rocks or schists.

Mica Schist.—This is, next to gneiss, the most abundant rock in New England. Specimen 43 is a typical example, and from it we can readily learn what mica schist is. A glance suffices to show that it is chiefly composed of mica, but not entirely; for, on carefully examining the edges of the specimen, we cannot fail to see thin layers of hard, glassy quartz interwoven with the mica. The quartz layers are short and overlapping, and we have here a good illustration of the schistose texture; this is, in fact, a typical schist.

Mica schist usually consists, as in this instance, of mica and quartz; but it may be composed of mica alone; and sometimes kaolin or clay takes the place of the quartz, forming argillaceous mica schist. The mica in the latter is usually in very fine scales and rather inconspicuous, and the rock often passes into ordinary clay slate. Similarly, when the mica becomes deficient in the quartzose mica schist, a passage into ordinary quartzite is the result. A little feldspar is sometimes present in the rock, which thus passes into micaceous gneiss. Specimen 43 contains several crystals of red garnet, giving the variety garnetiferous mica schist. There is no other rock that contains such a large variety of beautiful accessory minerals as mica schist; and for the mineralogist it is one of the most attractive rocks. Few rocks are more distinctly stratified; and the stratification can usually be observed in hand-specimens. The mica in these rocks may be either muscovite or biotite, or both; but the former is most common. No rock shows a greater variation in the percentage of silica which it contains than mica schist, as we pass from varieties which are nearly all quartz to those which are nearly all mica.

Closely related to mica schist is the rock now known as hydromica schist, in which the ordinary anhydrous micas are replaced by hydromica. It is distinguished from mica schist by being somewhat softer, less harsh to the touch, and less lustrous. It is to be regarded usually as an incipient mica schist, which has not yet become anhydrous; though it may sometimes be just the reverse; viz.: an old mica schist which has become hydrated through the action of meteoric waters. It contains fewer accessory minerals than mica schist.

Hornblende Schist.—This is a stratified aggregate of hornblende and quartz. The quartz is granular and in thin layers, as in mica schist; but the micaceous structure is wanting, and consequently the rock does not cleave readily in the direction of the bedding. The hornblende is mostly finely crystalline, but sometimes occurs in large, bladed crystals. Garnet and some other minerals are of common occurrence in the rock; but it is not rich in accessories like mica schist. The chief difficulty in recognizing this rock consists in determining whether the white mineral is all quartz or partly feldspar. In the latter case, of course, it becomes a hornblendic gneiss.

Amphibolite (Hornblende Rock).—This is the name applied to a rock having hornblende as its sole essential constituent. Hornblende schist sometimes passes into amphibolite, through the absence of quartz; and so does diorite, when the feldspar is deficient or wanting. Specimen 20, though small, is a typical example of this rock. The physical and chemical characteristics are essentially the same as for the mineral hornblende. The texture varies from coarsely to finely crystalline. The crystals are usually short and thick, and lie in all directions in the rock, which is thus very massive, the schistose texture being entirely wanting, and the stratification rarely showing in small masses. Biotite is a common accessory in amphibolite, and garnet and magnetite frequently occur.

By the substitution of augite for hornblende, in the description of amphibolite, we get the much rarer, but otherwise very similar, rock, *pyroxenite*.

Talc Schist (Steatite or **Soapstone**).—Although not abundant, this is a useful and familiar rock. The composition is implied in the name; and by comparing it with the specimen of talc (No. 58) we can readily see that they are essentially identical. Typical talc schist is pure talc; but the talc is often mixed with more or less quartz or feldspar; and mica, chlorite, hornblende, garnet, and other minerals are of common occurrence.

This rock embraces two distinct varieties, the massive and the schistose, or foliated. The former is the common soapstone (specimen 71), which is a confused mass of crystals lying in all directions, and with no visible stratification in the small mass. In the latter, as in specimen , the talc scales lie in parallel planes, giving the rock a micaceous structure, and causing it to split easily in the direction of the stratification. The cleavage surfaces are often wavy or corrugated; and the same is true of all schistose rocks. Talc schist is easily distinguished from all other rocks by its light-grayish or greenish color, combined with its extreme softness, and its smooth, slippery feel.

Chlorite Schist.—The one essential constituent of this rock is chlorite, and the mineral specimen (No. 26) answers equally well as an example of the rock. As with talc schist, quartz, feldspar, and hydromica are rarely entirely absent. Besides these, the principal accessories are hornblende, magnetite, garnet, and epidote. This rock also agrees with talc schist in presenting two principal varieties, the massive and the schistose. It is easily distinguished from talc schist by its darker color and streak, which are very characteristic; while its green color, softness, and unctuous feel separate it from all other rocks.

This is the most basic of all the silicate rocks; but, in consequence of containing a large proportion of water, it is not the heaviest. It is, in fact, interesting and important to observe that all these hydrous silicate rocks—talc schist, chlorite schist, greensand, and serpentine—are distinctly lighter in each case than anhydrous rocks containing the same proportion of silica. They are also notable, as a class, for their softness, smooth feel, and green color.

Serpentine.—As the name implies, this rock is simply the mineral serpentine occurring in large masses, and its characteristics are precisely the same. It is fine-grained, massive, compact, rather soft, but very tough, and varies in color from very dark green to light greenish-yellow. The dark colors predominate, and specimen 25 is a typical example.

Serpentine is often intimately associated with limestone and dolomite. The white veins running irregularly through the variety known as Verd Antique Marble, however, are not calcite, as commonly supposed, but magnesite. They do not effervesce freely with cold, dilute acid, for the entire rock is magnesian, and it is probable have been at one time simply cracks along which water holding carbon dioxide has penetrated, changing the magnesia from a silicate to a carbonate.

Geologists were, at one time, almost unanimous in the opinion that all serpentine is of eruptive origin; but now it is conceded by the great majority to be in some cases a sedimentary rock. It is found interstratified with gneiss, limestone, all the schists, and many other stratified rocks. When occupying the position of an eruptive it is never an original rock; but has been formed by the alteration, *in situ*, of some basic anhydrous rock, most commonly olivine basalt.

Greensand.—This rock (specimen 27) consists chiefly of the mineral glauconite, mingled usually with more or less sand, clay, or calcareous matter. It is usually very friable, or in an entirely unconsolidated state. It is most abundant in the newer geological formations, especially the Cretaceous and Tertiary; and is, perhaps, the only one of the stratified silicate rocks now forming on an extensive scale in the ocean. Its value as a fertilizer, for which purpose it is extensively employed, is due to the potash that it contains.

Following is a systematic summary of the mineralogical composition of the rocks of this great division of silicates; and this, combined with the classification on page 69, presents in a condensed form all the more important facts contained in the preceding descriptions. Only the more constant and normal constituents of the species are enumerated in each case:—

Names of Species.	Constituent Minerals.
Gneiss	Orthoclase and Quartz. Orthoclase, Quartz, and Mica. Orthoclase, Quartz, and Hornblende.
Syenite	Orthoclase. Orthoclase and Hornblende. Orthoclase and Mica.
Diorite	Plagioclase (chiefly Oligoclase). Plagioclase and Hornblende. Plagioclase and Mica.
Norite	Plagioclase (chiefly Labradorite). Plagioclase and Augite (Diallage). Plagioclase and Mica.

Mica Schist	Mica. Mica and Quartz. Mica and Kaolin.
Hornblende Schist	Hornblende and Quartz.
Amphibolite	Hornblende.
Pyroxenite	Pyroxene.
Talc Schist	Talc.
Chlorite Schist	Chlorite.
Serpentine	Serpentine.
Greensand	Glauconite.

2. Eruptive or Unstratified Rocks.

The rocks of this great class are formed by the cooling and solidification of materials that have come up from a great depth in the earth's crust in a melted and highly heated condition. When the fissures in the earth's crust reach down to the great reservoirs of liquid rock, and the latter wells up and overflows on the surface, forming a volcano, then we may, as was pointed out on page 33, divide the eruptive mass into two parts: first, that which has actually flowed out on the surface, and cooled and solidified in contact with the air, forming a lava flow; second, that which has failed to reach the surface, but cooled and solidified in the fissure, forming a dike.

Lava flows or volcanic rocks and dikes or plutonic rocks are identical in composition; but there is a vast difference in texture, due to the widely different conditions under which the rocks have solidified. The dike or fissure rocks solidify under enormous pressure, and this makes them heavy and solid—free from pores. They are surrounded on all sides by warm rocks: this causes them to cool very slowly, and allows the various minerals time to crystallize. Other things being equal, the slower the cooling the coarser the crystallization; and hence, the greater the depth below the surface at which the cooling takes place, the coarser the crystallization.

The volcanic rock, on the other hand, cools under very slight pressure; and the steam, which exists abundantly in nearly all igneous rocks at the time of their eruption, is able to expand, forming innumerable small vesicles or bubbles in the liquid lava; and these remain when it has become solid. Cooling in contact with the air, the lava cools quickly, and has but little chance for crystallization. Hence, to sum up the matter, we say: plutonic rocks are solid and crystalline; and volcanic rocks are usually porous or vesicular, and uncrystalline.

As we descend into the earth's crust, it is perfectly manifest that the volcanic must shade off insensibly into the dike rocks, and we find it impossible to draw any but an arbitrary plane of division between them; but this is no argument against this classification, for, as already stated, all is gradation in geology, and we experience just the same difficulty in drawing a line between conglomerate and sandstone, or between gneiss and mica schist, as between the dike rocks and volcanic rocks.

We will now observe to what extent the distinctions between these two great classes of eruptives can be traced in the rocks themselves, beginning with the dike rocks. But first it is important to notice the general fact, clearly expressed in the classification, that, with perhaps some trifling exceptions which need not be mentioned here, all eruptive rocks are silicates, and nearly all are feldspathic silicates. They are of definite mineralogical composition, and, like the chemically and organically formed stratified rocks, can be classified chemically. But, although there are eruptives corresponding closely in composition to the feldspathic silicates, which we have just studied, we find among them little to represent the non-feldspathic silicates, and nothing corresponding in composition to the limestones, dolomites, gypsum, flint, tripolite, siliceous tufa, iron-ores, bitumens or coals.

1. Plutonic (Dike) Rocks. — These are also known as the *ancient* eruptive rocks, and for this reason: It is impossible, of course, for us to observe them except where they occur on or near the earth's surface. But, since they are formed wholly below the surface, and usually at great depths in the earth, it is evident that they can appear on the surface only as the result of enormous erosion; and erosion is a slow process, demanding, in these cases, many thousands or millions of years. Therefore, the more ancient dike rocks alone are within our reach; those of recent formation being still deeply buried in the earth's crust. It follows, as a corollary to this explanation, that the coarseness of the crystallization of any dike rock must be a rough measure of its age and of the amount of erosion which the region has suffered since its eruption.

As regards composition, the dike rocks present, as already stated, essentially the same combinations of minerals as the feldspathic silicates of the stratified series, but occurring under different physical conditions and having a widely different origin. The only important difference in texture between the two classes of rocks is that the sedimentary rocks are stratified and the dike rocks are not; and when we consider that the dike rocks sometimes present a laminated structure that resembles stratification, while the sedimentary rocks frequently appear unstratified, it is easy to understand why, in the absence of any marked difference in composition, geologists have often found it difficult to distinguish the two classes of rocks. We also find here the explanation and the justification of the fact that the names of the dike rocks are in most cases the same as those of the sedimentary rocks of similar composition.

Granite. — Granite (from the Latin *granum*, a grain) is a crystalline-granular rock, agreeing in composition with gneiss. The essential constituents are quartz and orthoclase; and when they alone are present we have the variety *binary granite*. Mica, however (commonly muscovite, sometimes biotite, and frequently both) is usually added to these, forming *micaceous granite* (specimen 44); and often hornblende, forming *hornblendic granite* (specimen 45). The orthoclase is sometimes replaced in part by triclinic species, especially albite and oligoclase. Accessory minerals are not so abundant in granite as in gneiss; but, besides those named, garnet, tourmaline, pyrite, apatite, and chlorite are most common. Orthoclase is always the predominant ingredient; and, except when there is much hornblende present, usually determines the color of the granite. Thus, specimens 44 and 45 are gray because they contain gray orthoclase; while all red granites contain red or pink orthoclase. The quartz has usually been the last of the constituents to crystallize or solidify; and, having been thus obliged to adapt itself to the contours of the orthoclase and mica, it is rarely observed in distinct crystals.

In texture, the granites vary from perfectly compact varieties, approaching petrosilex, to those which are so coarsely crystalline that single crystals of orthoclase measure several inches in length. Of course one of the most important things to be observed about granite, especially in comparing it with gneiss, is the complete absence of anything like stratification; that, as before stated, being the only important distinction between the two rocks. Gneiss is the most abundant of all stratified rocks, and granite stands in the same relation to the eruptive series.

Syenite.—This is an instance where stratified and eruptive rocks, agreeing in composition, have the same name. That rocks consisting of orthoclase, of orthoclase and hornblende, or of orthoclase and mica, *i.e.*, having the composition of syenite, do occur in both the eruptive and stratified series there can be no doubt. They should, however, have distinct names on account of their unlike origins; and would have but for the practical difficulty in determining, in many cases, whether the rock is stratified or not. The best that we can do now, when we desire to be specific, and have the necessary information, is to say stratified syenite or eruptive syenite, as the case may be.

Diorite.—Here, again, we find identity of names, as well as of composition, between the two great series. Eruptive diorite is an abundant and well known rock, and consists of the same minerals as stratified diorite combined in the same proportions. Diorite includes a large part of the dike rocks commonly known as "trap" and "greenstone." The principal accessories are chlorite, epidote, pyrite, magnetite, apatite, and quartz. The texture varies from perfectly compact or felsitic to coarsely crystalline; averaging, however, less coarse than syenite and granite.

Diabase.—By referring to the classification it will be seen that diabase occupies the same position among the dike rocks as norite among the stratified rocks. Like norite it consists usually of the more basic varieties of plagioclase with or without augite, diallage, or hypersthene. Augite, or one of its representatives, is usually present, and is often the principal constituent. Specimen 1 shows a somewhat equal development of the feldspar and augite. The name *gabbro* is sometimes applied to the coarser and more feldspathic diabases, and especially to those containing diallage or hypersthene in the place of common augite. In the opinion of some high authorities, however, it is unnecessary to recognize two species here; and it makes the classification more simple and symmetrical not to do it. The principal accessories in diabase are biotite, chlorite, magnetite, pyrite, calcite, and olivine. Chlorite is often an important constituent, giving the rock a greenish aspect; but here, as well as in diorite, the chlorite is due chiefly or entirely to the alteration of the augite and feldspar; and the chloritic varieties of diorite and diabase together make up the old species "greenstone." Similarly, the more compact and darker varieties of these two rocks, forming regular, wall-like dikes, are known as "trap."

Specimen 46.

In consequence of their more basic composition, diabase and diorite are usually strongly contrasted with granite and syenite in color and specific gravity, being darker and heavier. The basic rocks, too, decay much more readily than the acidic.

2. VOLCANIC ROCKS.—As regards composition, we shall find nothing new in the volcanic series; for the rocks of this group present essentially the same combination of minerals as the dike rocks. In composition, the dike and volcanic rocks are identical; but in texture, as already explained, there is a vast difference. The volcanic rocks differ so widely in texture from both the dike and stratified species, that there is rarely any difficulty in distinguishing them; and hence they have in every instance distinct names.

Volcanic rocks are rarely found in this part of the world; and specimens of most of them are difficult to obtain. For this reason they can only be noticed briefly here, since it is the plan of this Guide to give especial attention only to those portions of the subject which can be illustrated by material within easy reach of teachers.

Rhyolite.—This rock corresponds in composition with granite and gneiss, but is less frequently micaceous. The orthoclase in rhyolite, and generally in volcanic rocks, is the clear, pellucid variety—*sanidine*. It is more difficult to separate from quartz than ordinary orthoclase, the chief distinguishing feature being its cleavage. Plagioclase and hornblende are common, but not abundant, constituents. The mica, when present, is usually biotite. The texture of rhyolite is often more or less distinctly porphyritic, having a finely crystalline or granular matrix, with interspersed crystals of sanidine and quartz. The rock has usually a rough, harsh feel; and while the coarser varieties have the aspect of granite, the finer approach petrosilex; but all are somewhat porous, which is seen in the lower specific gravity of rhyolite as compared with granite and gneiss.

Trachyte.—In texture and general aspect rhyolite and trachyte are nearly identical. Trachyte, however, is darker, contains little or no quartz, and more hornblende and plagioclase. In fact, it agrees in composition with syenite. This is one of the most important of the volcanic rocks.

Obsidian.—Obsidian is sharply distinguished from all other rocks by its perfect vitreous texture; it is a true volcanic glass. Its surface (specimen 47) is smooth and glassy, and its fracture eminently conchoidal. To the naked eye, and usually under the microscope, the typical variety is perfectly homogeneous; chemical analysis, however, shows that it has the composition, commonly of rhyolite, but sometimes of trachyte. Obsidian is, in fact, simply rhyolite or trachyte which, cooling quickly, has not had time to crystallize, but has remained permanently in the amorphous or glassy state. The composition is sometimes partially revealed where a portion of the sanidine comes out in distinct crystals porphyritically interspersed through the glass. The homogeneity of the texture is sometimes disturbed: by numerous minute concentric cracks, forming what is known as perlitic structure and the variety perlite; by numerous small spherical concretions, forming the spherulitic structure and the variety spherulite; and also by the banding, which is the result of flowing while in a plastic state, whereby portions of the glass of slightly different colors are drawn out into layers and interlaminated. The bands are rarely continuous for any distance, being usually merely elongated lenticular streaks. The glassy state is generally one of inferior density, and hence we find that obsidian is lighter than the crystalline rocks of the same composition. Obsidian is a good illustration of a non-essential color, for its capacity and jet-black color are due entirely to impurities. In very thin flakes it is transparent and white. It also forms a white powder when crushed, *i.e.*, it has a white streak.

Obsidian is often vesicular, from the expansion of the steam and other gases which it contained when liquid. The most thoroughly vesicular varieties are known as *pumice* (specimen 48). The vesicular texture, by rendering the rock impervious to light, conceals the impurities, and thus we get a snow-white pumice from black obsidian. The vesicles are frequently elongated, sometimes in a definite direction, though often forming an irregular net-work of glassy fibres. Pumice is often light enough to float on water, and it is transported thousands of miles by the oceanic currents. It is employed in the arts, and good specimens can be obtained at almost any drug-store.

Petrosilex and **Felsite**.—Sharply defined groups are unknown in lithology, but all is gradation; and between rhyolite and trachyte, which are always more or less distinctly crystalline, and obsidian, which is a true glass and perfectly amorphous, there is no break. It is impossible to draw a sharp line and say, Here the vitreous texture ends and the crystalline begins; for the transition is not abrupt, but gradual. We recognize, really, in these feldspathic rocks, an intermediate state, which is neither crystalline nor colloid, but both; and this lithologists have designated the *felsitic* texture. Felsitic matter cannot, even with the highest powers of the microscope, be resolved into separate grains or particles; and it does not exhibit, except perhaps very indistinctly, the phenomenon of double refraction. In other words, it is not truly crystalline or stony, and yet it is just as clearly not amorphous or glassy.

Feldspathic rocks exhibiting the felsitic texture in whole or in part are known as *felsites*. Many high authorities hold that true felsites are found only among the eruptive rocks; while others claim that they are in part, or wholly, of sedimentary origin. The writer accepts the former view. The felsites are in part acid lavas which have cooled too slowly to form a true glass, like obsidian, and yet too quickly to become truly crystalline, like rhyolite and trachyte. But they are also in large part simply devitrified obsidian. Glass is an unstable form of mineral matter; and every species of glass, including obsidian, tends with the lapse of time to become crystalline or stony, the amorphous changing to the felsitic structure. Thus, in many cases or usually, what we now call felsites were originally true glassy obsidian. Being perfectly intimate mixtures of the component minerals, the composition of felsites can usually be determined with certainty only by means of chemical analysis. By this means chiefly, it has been proved that there are felsites agreeing in composition with both rhyolite and trachyte. There is this general difference in composition, however, between these crystalline rocks and the felsites; viz.: mica, hornblende, and augite are generally wanting in the latter. From this it follows that the felsites are, with unimportant exceptions, composed either of quartz and feldspar or of feldspar alone.

The physical differences between the felsites of unlike composition are not great; but they are sufficient to warrant the division of the felsites into two species: a basic species, to which the term *felsites* may properly be restricted; and an acidic species, for which *petrosilex* is a very appropriate name. According to this arrangement, felsite is composed chiefly of orthoclase, and, as the table shows, agrees in composition with trachyte; while petrosilex consists mainly of orthoclase and quartz, agreeing in composition with rhyolite. We find here nothing new in composition; but petrosilex and felsite are simply the crystalline rocks which we have already studied, repeated under a different texture.

The typical felsite or petrosilex is composed entirely of felsitic matter, and is perfectly homogeneous, like flint or jasper, which it closely resembles in hardness and other physical characteristics. As a rule, however, the rock is not entirely homogeneous, but there is a manifest tendency in the component minerals, and especially in the feldspar, to separate out, usually in the form of crystals. In the banded variety (specimen 42) the rock is built up of thin layers, which are often alternately quartzose and feldspathic. There is not a perfect separation of the minerals; but that the quartz is chiefly in the dark layers, and the feldspar in the light, is shown by the way in which the layers are affected by the weather.

One of the most common varieties is where a portion, frequently a large portion, of the feldspar comes out in the form of distinct, separate crystals, producing a porphyritic texture. Specimens 5, 6, and 7 are examples of porphyritic felsite; and after examining these we can no longer doubt that feldspar is an important constituent of the rock. Petrosilex and felsite are more generally porphyritic than any other rocks; and they are commonly called porphyry. It is better, however, since almost any rock may be porphyritic, and since this texture cannot be correlated with any particular composition, not to use porphyry as a rock-name, but simply as the name of a very important rock-texture. The banded and porphyritic textures are about equally characteristic of petrosilex and felsite. In petrosilex, quartz, as well as feldspar, is sometimes porphyritically developed, forming the variety known as quartz-porphyry. There is no limit to the proportion of the quartz and feldspar which may crystallize out in this way, and thus we find a perfectly gradual passage from normal petrosilex or felsite to thoroughly crystalline granite and syenite.

Andesite.—This rock has nearly the texture of rhyolite and trachyte, but is darker and heavier, and corresponds in composition to diorite, consisting of plagioclase and hornblende, with usually more or less sanidine, quartz, augite, biotite, and magnetite.

Basalt.—The rock bearing this familiar name represents diabase among the dike rocks. It is the most basic of the volcanic rocks, and consists of the more basic varieties of plagioclase, especially labradorite, with augite, magnetite, and titanic iron. Olivine is a very common and characteristic constituent, and the plagioclase is often replaced in part by leucite and nephelite. The basalts are usually black, and of high specific gravity; and vary in texture from compact to coarsely crystalline. The contraction due to cooling frequently results in the development of a columnar structure of remarkable regularity, the columns being normally hexagonal and standing perpendicularly to the cooling surfaces of the mass. This structure occurs in other eruptive rocks, but is most characteristic of basalt.

Tachylite.—Tachylite is a highly basic volcanic glass, standing in the same relation to basalt and andesite that obsidian does to trachyte and rhyolite. It is much heavier than obsidian, and is perfectly black and opaque, except in the finest fibres. It is a comparatively rare rock, for the reason that basalt and andesite crystallize more readily than the acidic rocks on passing from the liquid to the solid state. On the surface of the basic lava, however, where it is in contact with the air, and congeals almost instantly, a film of glass is formed; but this may not be more than a small fraction of an inch in thickness. Like obsidian, tachylite is often vesicular; but the vesicular basic rocks, as well as the solid, are usually stony. They occur in vast abundance in many volcanic regions, and may be considered the typical lava (specimen 49).

In the more ancient lavas, the vesicles are frequently filled by various minerals—chlorite, epidote, quartz, calcite, etc.—deposited by infiltrating waters, and derived in most cases from the decomposition of the original constituents of the rock. Thus the vesicular is changed to the amygdaloidal texture, and the lava becomes an amygdaloid (specimen 50). The amygdaloidal texture is common in the basic lavas and rare in pumice, simply because the former are more readily decomposed and contain a greater variety of bases from which secondary minerals can be formed.

Porphyrite and **Melaphyr.**—These two rocks hold essentially the same relation as regards origin and structure to the basic lavas that petrosilex and felsite do to the acidic lavas. Porphyrite agrees in composition with andesite, and melaphyr with basalt. They are usually dark-colored rocks having a compact or felsitic texture. Porphyrite is, as the name implies, very commonly porphyritic; while melaphyr is often vesicular or brecciated, exhibiting all the structural features of tachylite and basalt, and being in its older forms very generally amygdaloidal.

Volcanic Tuff and **Agglomerate.**—Besides the crystalline, glassy, and felsitic lavas, already described, and due chiefly to the rate of cooling of the liquid rock, we may recognize a fourth class to include the very abundant lavas which, during explosive eruptions, are ejected in the solid state, being violently blown out of the crater in the form of dust and fragments. Falling on the slopes of the volcano or over the surrounding country, as in the case of the buried city of Pompeii, the fragmental lavas remain largely unstratified. But when, as frequently happens, they fall into the sea, they are assorted by the waves and currents and arranged in layers after the manner of ordinary sediments, with which they are often more or less mixed. Before they become consolidated the finer fragmental lava, of whatever composition, is called volcanic dust, and the coarser lapilli or volcanic sand; while the consolidated materials are known as tuff and agglomerate respectively.

SUPPLEMENT TO LITHOLOGY.

VEIN ROCKS.

All rocks are not embraced in the sedimentary and eruptive divisions, but there is a third grand division, which, although rarely mentioned or recognized in the more comprehensive works on geology, it is deemed best not to leave entirely unnoticed here. These are the *vein* rocks. They present an immense number of varieties, and yet, taken altogether, form but a small fraction of the earth's crust. They are, however, the great repositories of the precious and other metals, and hence are objects of far greater interest to the miner and practical man than the eruptive rocks, or, in some parts of the world, even than the sedimentary rocks.

The vein rocks, like the eruptive rocks, occupy fissures in the earth's crust intersecting the stratified formations; but the fissures filled with vein rocks are called veins, and not dikes. We will first notice the mode of formation of a typical vein, and then examine its contents. Geologists are agreed that water penetrates to a very great depth in the earth's crust. All minerals are more or less soluble in water; and we may consider the water circulating through the rocks, especially at considerable depths, as, in most cases, a saturated solution of the various minerals of which they are composed. Very slight changes in the conditions will cause saturated solutions to deposit part of their mineral load. The water at great depths has a high temperature, and is subjected to an enormous pressure; and both of these circumstances favor solution. Suppose, now, that these hot subterranean waters enter a fissure in the crust and flow upwards, perhaps issuing on the surface as a warm mineral spring; as they approach the surface, the temperature and pressure, and consequently their solvent power, are diminished; and a portion of the dissolved minerals must be deposited on the walls of the fissure, which thus becomes narrower, and in the course of time is gradually filled up. The vein is then complete; and the mineral waters are forced to seek a new outlet.

Veins have the same general forms as dikes, since the fissures are formed in the same way for both; but the vein is of slow growth, and may require ages for its completion, while the dike is formed in an hour or a day. It is now generally believed that water is an important agent in the formation of eruptive rocks; since they all contain water at the time of their eruption; and since it has been demonstrated that, while ordinary rocks require a temperature of 2000° to 3000° for their fusion in the absence of water, they are liquified at temperatures below 1000° in the presence of water. In other words, common rocks are very infusible and insoluble bodies, and heat and water acting independently have little effect upon them; but when fire and water are combined in what is now known as aqueo-igneous fusion, they prove very efficient agents of liquefaction.

If we adopt these views, then it can be shown that, in origin, veins and dikes differ in degree only, and are not fundamentally unlike; and the formation and relations of the three great classes of rocks may be summarized as follows:—

The ocean and atmosphere, operating on the earth's surface, have worked over and stratified the crust, until the sedimentary rocks have now an average thickness variously estimated at from ten to thirty or forty miles. This entire thickness of stratified rocks, and a considerable depth of the underlying unstratified crust, must be saturated with water; and all but the more superficial portions of this water-soaked crust must be very hot, the temperature increasing steadily downwards from the surface. Both eruptive and vein rocks originate in this highly heated, hydrated crust. Eruptive rocks are formed when the heat, aided by more or less water, softens the rocks, either stratified or unstratified, by aqueo-igneous fusion, and the plastic materials are forced up through fissures to or toward the surface. Vein rocks are formed when the water, aided by more or less heat, dissolves the rocks, either stratified or unstratified, by what may be called igneo-aqueous solution, and subsequently deposits the mineral matter in, *i.e.*, on the walls of, fissures leading up to or toward the surface. In the case of the dike rocks, heat is the chief agent, and water merely an auxiliary; while with the vein rocks it is just the reverse. But between the two it is probably impossible to draw any sharp line.

The water circulating through the crust, and saturated with its various mineral constituents, has been called the "juice" of the crust; and veins are formed by the concentration of this earth-juice in fissures. One of the most important characteristics of the vein rocks, as a class, is the immense variety which they present; for nearly every known mineral is embraced among their constituents; and these are combined in all possible ways and proportions, so that the number of combinations is almost endless. The solvent power of the subterranean waters varies for different minerals; and appears often to be greatest for the rarer species. In other words, there is a sort of selective action, whereby many minerals which exist in stratified and eruptive rocks, so thinly diffused as to entirely escape the most refined observation, are concentrated in veins in masses of sensible size; and our lists of known minerals and chemical elements are undoubtedly much longer than they would be if these wonderful storehouses of fine minerals which we call veins had never been explored. As a rule, the minerals in veins form larger and more perfect crystals than we find in either of the other great classes of rocks. This is simply because the conditions are more favorable for crystallization in veins than in dikes or sedimentary strata. In both dike and stratified rocks, the growing crystals are surrounded on all sides by solid or semi-solid matter; and, being thus hampered, it is simply impossible that they should become either large or perfect. In the vein, on the other hand, there are usually no such obstacles to be overcome; but the crystals, starting from the walls of the fissure, grow toward its centre, their growing ends projecting into a free space, where they have freedom to develop their normal forms and to attain a size limited only, in many cases, by the breadth of the fissure. With, possibly, some rare exceptions, all the large and perfect crystals of quartz, feldspar, mica, beryl, apatite, fluorite, and of minerals generally, which we see in mineralogical cabinets, have originated in veins. Those fissures which become the seats of mineral veins are really Nature's laboratories for the production of rare and beautiful mineral specimens; and hence the vein rocks are the chief resort of the mineralogist, to whom they are of far greater interest than all the eruptive and stratified rocks combined.

The leading characteristics, then, of the vein rocks may be summarized as follows: (1) They contain nearly all known minerals, including many rare species and elements which are unknown outside of this class of rocks. (2) These mineral constituents, occurring singly and combined, give rise to a number of varieties of rocks so vast as to baffle detailed description. (3) They exceed all other rocks in the coarseness of their crystallization, and in the perfection and beauty of the single crystals which they afford.

PETROLOGY.

In lithology we investigate the nature of the materials composing the earth's crust—the various minerals and aggregates of minerals, or rocks; while in petrology we consider the forms and modes of arrangement of the rock-masses,—in other words, the architecture of the earth.

Petrology is the complement of lithology, and in many respects it is the most fascinating division of geology, since in no other direction in this science are we brought constantly into such intimate relations with the beautiful and sublime in nature. The structures of rocks are the basis of nearly all natural scenery; for what we call scenery is usually merely the external expression, as developed by the powerful but delicate sculpture of the agents of erosion—rain and frost, rivers and glaciers, etc.—of the geological structure of the country. And to the practised eye of the geologist, a fine landscape is not simply a pleasantly or grandly diversified *surface*, but it has *depth*; for he reads in the superficial lineaments the structure of the rocks out of which they are carved.

But, while the magnitude of the phenomena adds greatly to the charm of the study, it also increases the difficulties and taxes the ingenuity of the teacher whose work must be done indoors. According to our ideal method, natural science ought to be taught with natural specimens; and yet here our main reliance must be upon pictures and diagrams.

Nature, however, has not been wholly unmindful of our needs; for she has worked often upon a very small as well as a very large scale; many of the grandest phenomena being repeated in miniature. Thus we observe rock-folds or arches miles in breadth and forming mountain masses, and of all sizes from that down to the minutest wrinkle. So with veins, faults, etc. And the wonderful thing is that these small examples, which may be brought into the class-room, are usually, except in size, exactly like the large. Now the aim of every teacher in this department should be to secure a collection of these natural models. It is not an easy thing to do, except one has plenty of time; for they can rarely be purchased of dealers, but must usually come as the choicest fruit of repeated excursions to the natural ledges and quarries, the seashore and the mountains. But for the difficulty of getting the specimens there is some compensation, since it may be truly said that for the *collector* specimens obtained in this way have an interest, a value, and a power of instruction beyond what they would otherwise possess.

Classification of Structures.

The structures of rock divide, at the outset, into two classes:—(1) the *original structures*, or those produced at the same time and by the same forces as the rocks themselves, and which are, therefore, peculiar to the class of rocks in which they occur (*e.g.* stratification, ripple-marks, fossils, etc.); and (2) the *subsequent structures*, or those developed in rocks subsequently to their formation, and by forces that act more or less uniformly upon all classes of rocks, and which are, therefore, in a large degree, common to all kinds of rocks (*e.g.* folds, faults, joints, etc.).

The original structures are conveniently and naturally classified in accordance with the three great classes of rocks: (1) stratified rocks, (2) eruptive rocks, and (3) vein rocks; while the subsequent structures, not being peculiar to particular classes of rocks, are properly divided into those produced by (1) the subterranean or igneous agencies, and (2) the superficial or aqueous agencies.

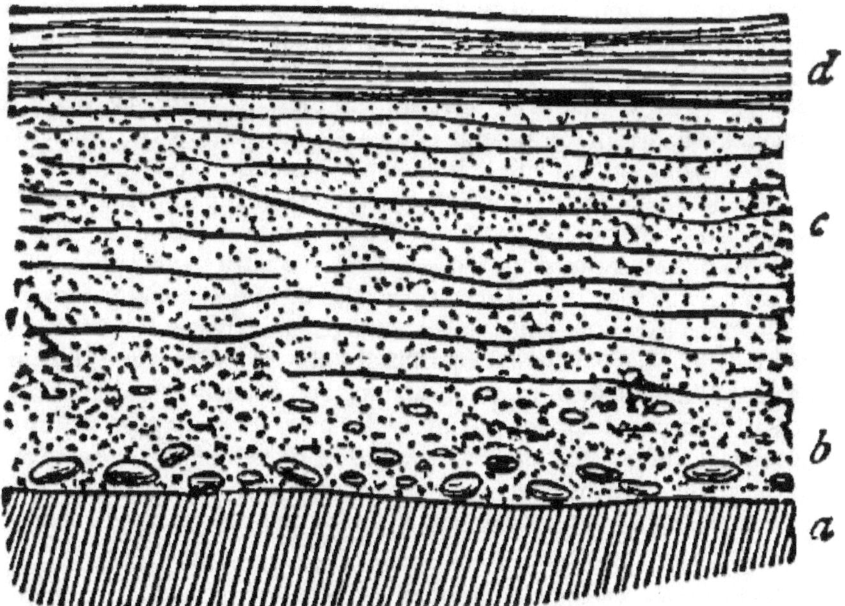

Fig. 1.—Section through sediment deposited by rain in a roadside pool:
a. surface of roadway; *b.* layer of small pebbles and coarse sand;
c. fine sand passing into *d*; *d.* the finest sand and mud.

Original Structures of Stratified Rocks.

STRATIFICATION.—All rocks formed by strewing materials in water, and their deposition in successive, parallel, horizontal layers, are *stratified*; and this structure is their *stratification*. It is the most important of all rock structures; and there is no kind of structure the origin of which is more fully or certainly known. The deposition of sediment in carefully assorted horizontal layers is readily brought within the comprehension of children by simple experiments with sand and clay in water; and still better by the examination of the deposits formed in roadside pools during heavy rains (Fig. 1), and by digging into beaches and sandbars, which every child will recognize as formed of materials arranged by water. Great stress should be laid upon the fact that a lake like Erie or Champlain is simply a large pool with several more or less turbid streams flowing into it, while the single stream flowing out is clear, the sediment having evidently been deposited in the lake; and that every lake is, like the roadside pool, being gradually filled up with sedimentary or stratified rocks. But the ocean is a still larger pool, receiving mud and sand from many streams; and since we know that nothing escapes from the ocean but invisible vapor, it is plain that the mud and sand and all other kinds of sediment carried into the ocean must be deposited on its floor, and chiefly, as we have seen, on that part nearest the land. The consolidation of beaches, bars, and mud-flats is all that is necessary to convert them into stratified formations of conglomerate sandstone and slate.

Let us notice now, more particularly, the causes of visible stratification. As we can easily prove by an experiment with clay in a bottle of water, if the same kind of material is deposited continuously there will be no visible stratification in the deposit. It will be as truly stratified as any formation, but not visibly so; because there is nothing in the nature of the material or the way in which it is laid down to bring out distinct lines of stratification. Continuous and uniform deposition obtains very frequently in nature, but rarely continues long enough to permit the formation of thick beds or strata. Hence, while the stratification is almost always visible on the large surfaces of sandstone, slate, etc., exposed in quarries and railway cuttings, and may usually be seen in the quarried blocks, it is often not apparent in hand specimens, which may represent a single homogeneous layer. There is one important exception, and that is where the particles, although of the same kind, are flat or elongated. Pebbles of these forms are common on many beaches; and since they are necessarily arranged horizontally by the action of the water, they will, by their parallelism, make the stratification of the pudding-stone visible. The same result is accomplished still more distinctly by the mica scales, etc., in sandstone and slates, the leaves and flattened stems of vegetation in bituminous coal, and the flat shells in limestone.

In all other cases, visible stratification implies some change in the conditions; either the deposition was interrupted, or different kinds of material were deposited at different times. The first cause produces planes of easy splitting, or fissility, especially in fine-grained rocks, like shale. This shaly structure or lamination-cleavage may be due, in some cases, to pressure, but it is commonly understood to mean that each thin layer of clay became partially consolidated before the next one was deposited upon it, so that the two could not perfectly cohere. Parallel planes of easy splitting are, however, by themselves, of little value as indications of stratification, since the lamination-cleavage is not easily distinguished from slaty-cleavage (roofing slate) and parallel jointing, structures developed subsequently to the deposition of sediments and quite independent of the stratification. The second cause, or variations in the kind of sediment, gives alternating layers differing in color, texture, or composition, as is seen frequently in sandstone, slate, gneiss, etc.; and of all the indications of stratification these are the most important and reliable.

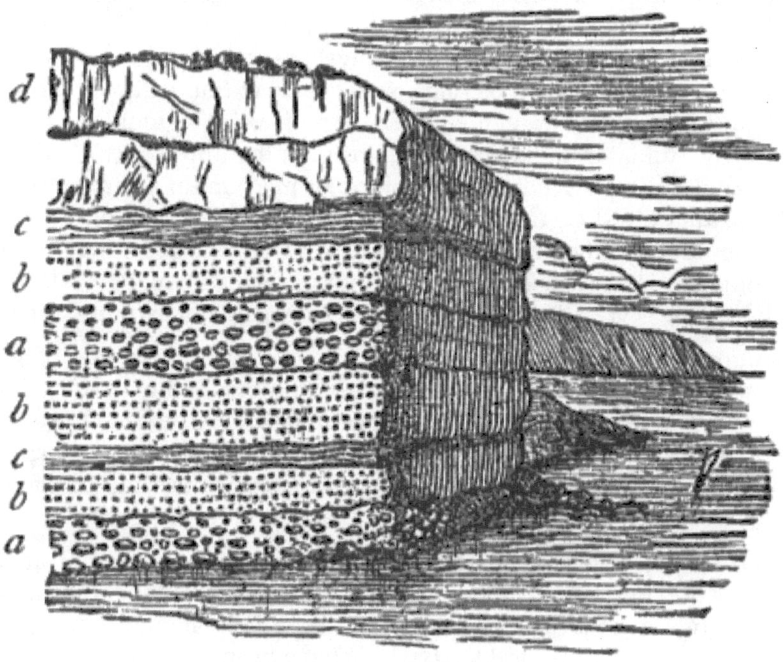

Fig. 2.—Section showing strata and laminæ: *a.* **conglomerate;** *b.* **sandstone;** *c.* **shale;** *d.* **limestone.**

A layer composed throughout of essentially the same kind of rock, as conglomerate or sandstone, and showing no marked planes of division, is usually regarded as one *bed* or *stratum*, although it may vary considerably in texture or color; while the thinner portions composing the stratum and differing slightly in color, texture, and composition, and the thin sheets into which shaly rocks split, are the *laminæ* or *leaves*. In Fig. 2 the strata are designated by letters, and the fine lines and rows of dots show the constituent laminæ, while the whole section may be regarded as a small part of a great geological formation. The geological record is written chiefly in the sedimentary rocks; and the formations, strata, and laminæ may be regarded as the volumes, chapters, and pages in the history of the earth. Now every feature of a rock, lithological or petrological, finds its highest interest in the light which it throws upon the history of the rock, *i.e.*, upon the conditions of its formation. Observe what the section in Fig. 2 teaches concerning the geological history of that locality; premising that any chapter of geological history written in the stratified rocks should be read from the bottom upwards, since the lowest strata must have been formed first and the highest last. The lowest stratum exposed is conglomerate, indicating a shingle beach swept by strong currents which carried away the finer material. Upwards, the conglomerate becomes finer and shades off into sandstone, and finally into shale, showing that the water has become gradually deeper and more tranquil, the shore having, in consequence of the subsidence, advanced toward the land. The next two strata show that this movement is probably reversed; at any rate, the currents become stronger again, and the shale passes gradually into sandstone and conglomerate. The beach condition prevails now for a long time, and thick beds of sand and gravel are formed. The sea then deepens again, and we observe a third passage from coarse to fine sediment. This locality is now remote from the shore, the gentle currents bringing only the finest mud, which slowly builds up the thick bed of shale, in the upper part of which shells are abundant, indicating that the deposition of mechanical sediment has almost ceased, and that the shale is changing to limestone. The purity of the limestone, and the crinoids and other marine organisms which it contains, prove that this has now become the deep, clear sea; and this condition is maintained for a long period, for the limestone is very thick, and this rock is formed with extreme slowness.

The most important point to be gained here is that every line of stratification and every change in the character of the sediments is due to some change of corresponding magnitude in the conditions under which the rock was formed. The slight and local changes in the conditions occur frequently and mark off the individual laminæ and strata, while the more important and wide-spread changes determine the boundaries of the groups of strata and the formations.

Strata are subject to constant lateral changes in texture and composition, *i.e.*, a bed or formation rarely holds the same lithological characteristics over an extended area. There are some striking exceptions, especially among the finer-grained rocks, like slate, limestone, and coal, which have been deposited under uniform conditions over wide areas. It is the general rule, however, particularly with the coarse-grained rocks, which have been deposited in shallow water near the land, that the same continuous stratum undergoes great changes in thickness and lithological character when followed horizontally. A stratum of conglomerate becomes finer grained and gradually changes into sandstone, which shades off imperceptibly into slate, and slate into limestone, etc. Where the stratum is conglomerate, its thickness will usually be much greater and more variable than where it is composed of the finer sediments. The rapidity of these changes in certain cases is well shown by the parallel sections in Fig. 3. These represent precisely the same beds, as the connecting lines indicate, at points only twenty feet apart.

When we glance at the conditions under which stratified rocks are now being formed, it is plain that all strata must terminate at the margin of the sea in which they were deposited, and in the marginal portions of that sea, especially, must exhibit frequent and rapid changes in composition, etc. The sediments forming the surface of the sea-bottom at the present time may be regarded as belonging to one continuous stratum; and it is instructive to examine a chart of any part of our coast, such as Massachusetts Bay, on which the nature of the bottom is indicated for each sounding, and observe the distribution of the different kinds of sediment. On an irregular coast like this, especially, the gravel, sand, and mud of different colors and textures, and the different kinds of shelly bottom, form a patchwork, the patches being, for the most part, of limited extent and shading off gradually into each other.

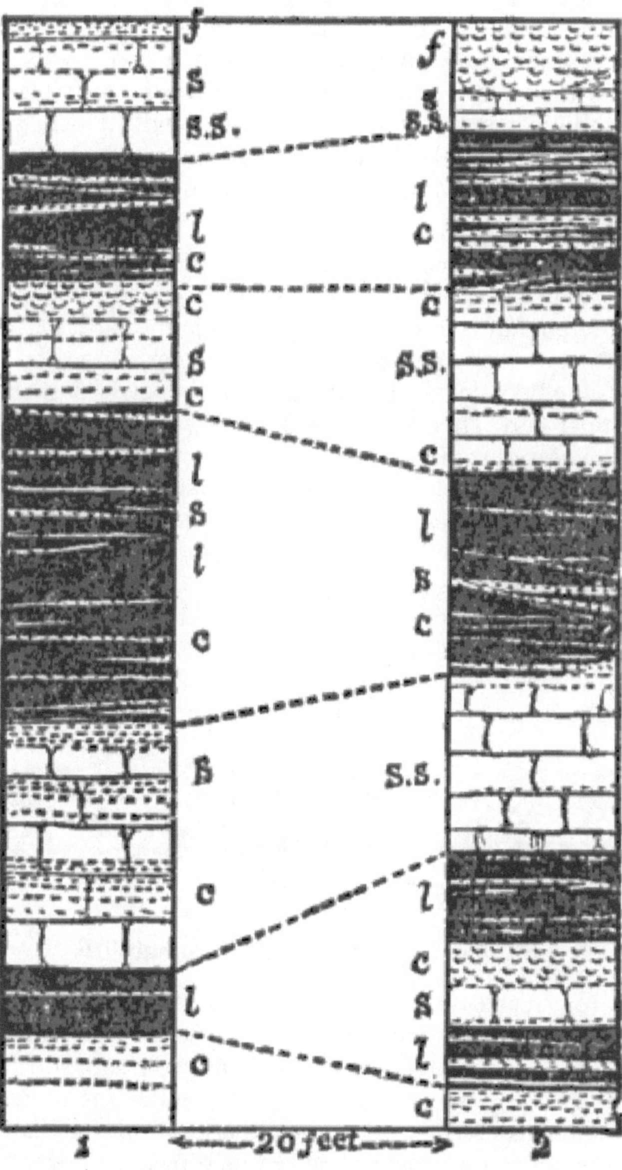

Fig. 3.—Parallel sections showing rapid lateral changes in strata: *c.* clay; *s.* sand; *ss.* sandstone; *l.* lignite; *f.* fireclay.

On a more regular coast, like that of New Jersey, the sediments are distributed with corresponding uniformity, the changes are less frequent and more gradual, and we have here a better chance to observe the normal arrangement of the sediments along a line from the shore seawards—gravel, sand, mud, and shells. On the beach we find the shingle and coarse pebbles, shading off rapidly into fine pebbles and sand. The zone or belt of sandy bottom may vary in width from a mile or two to twenty miles or more, becoming gradually finer and changing into clay or mud, which covers, usually, a much broader zone, sometimes extending into the deeper parts of the sea, but gradually giving way to calcareous sediments. Hence we may say that the finer the sediment the greater the area over which it is spread; but, on the other hand, the coarser the sediment the more rapidly it increases in thickness. In other words, the horizontal extent of a formation deposited in any given period of time is inversely, and the vertical extent or thickness is directly, proportional to the size of the particles.

Observations made in deep wells and mines, and where, by upturning and erosion, the edges of the strata are exposed on the surface, show that the vertical order of the different kinds of sedimentary rocks in the earth's crust is extremely variable. But when we take a general view of a great formation, it is often apparent that it consists chiefly of coarse-grained rocks in the lower part and fine-grained rocks in the upper part. This is, in general, a necessary consequence of the fact that a great thickness of sediments can only be formed on a subsiding sea-floor. Such a formation must consist chiefly of shore deposits, and be deposited near the shore where the sea is shallow. Hence, 10,000 feet of sediments implies nearly that amount of subsidence. In consequence, the shore line and the several zones of sediment advance towards the land; and sand is deposited where gravel was at first, and as the subsidence continues, both clay and limestone are finally deposited over the original beach. When the sea-floor rises, the order of the sediments is reversed; and it will be observed that in consequence of the advance and retreat of the shore-line, the formations grow edgewise to a considerable extent.

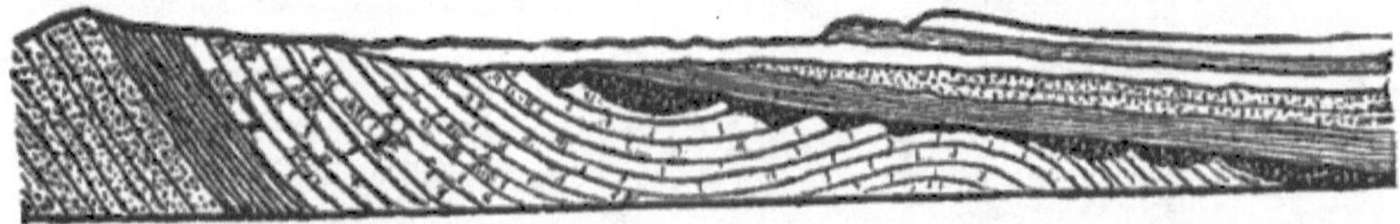

Fig. 4.—Overlap and unconformability.

Overlap and Interposition of Strata.—Another consequence of the constant oscillation of the shoreline is that successive deposits in the same sea will often cover different and unequal areas. When, in consequence of subsidence, one formation extends beyond and covers the edge of another, as shown in Fig. 4, we have the phenomenon described as overlap. Interposition is similar, being the case where a formation (Fig. 5, *c.*) does not, in certain directions, cover so wide an area as the strata (*b. d.*) above and below it, which are thus sometimes found in contact, although normally separated by the entire thickness of the intermediate and, seemingly, interposed stratum.

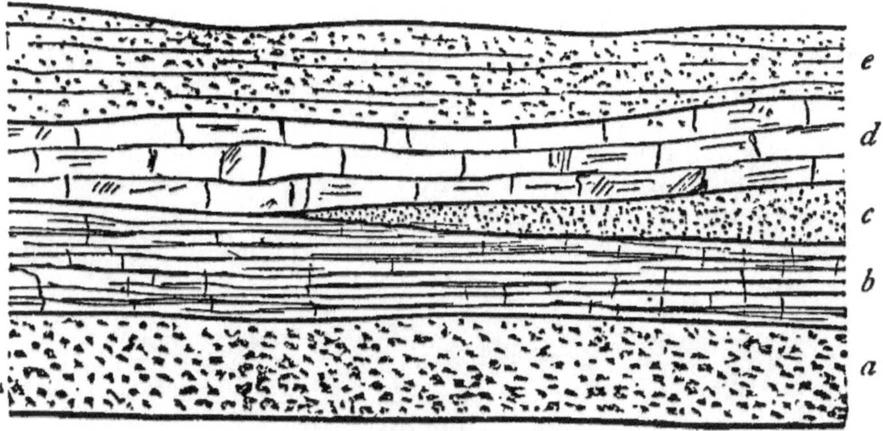

Fig. 5.—Interposition of strata.

UNCONFORMABILITY.—We have already seen that the rocks on the land are being constantly worn away by the agents of erosion; and it is also a matter of common observation that the strata thus exposed are often not horizontal, but highly inclined, having been greatly disturbed and crumpled during their elevation. Now, when such a land-surface subsides to form the sea-bottom, and new strata are spread horizontally over it, they will lie across the upturned and eroded edges of the older rocks, and fill the hollows worn out of the latter, as shown in Fig. 6; and the new formation is then said to rest unconformably upon the older. Two strata or formations are unconformable when the older has suffered erosion (Fig. 6), or both disturbance and erosion (Fig. 4) before the deposition of the newer.

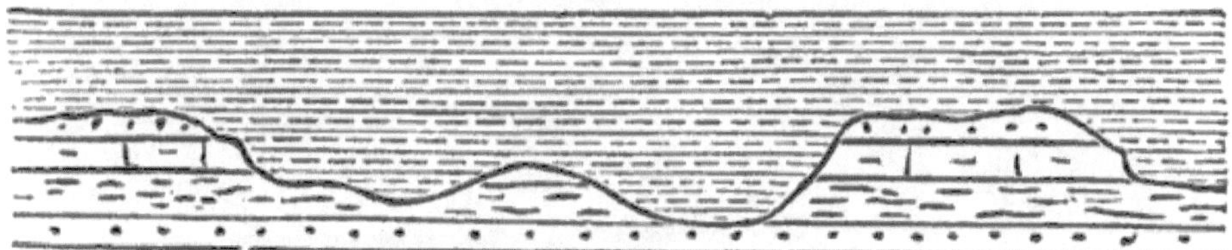

Fig. 6.—Unconformability.

When strata are conformable, the deposition may be presumed to have been nearly or quite continuous; but unconformability clearly proves a prolonged interruption of the deposition during which the elevation, erosion, and subsidence of the sea-bottom took place. The section in Fig. 7 shows a second unconformability, proving that the sea-bottom has here been lifted three times to form dry land. An unconformability may sometimes be clearly established when the actual contact of the two formations cannot be seen, as where the new formation is a conglomerate containing fragments of the older.

IRREGULARITIES OF STRATIFICATION.—These are especially noticeable in sandstone and conglomerate, which have been deposited chiefly by strong, local, and variable currents; the kind and quantity of sediment, of course, varying with the strength and direction of the current. Two kinds of irregularity only may be specially noticed here: (1) contemporaneous erosion and deposit, where, in consequence of a change in the currents, fine material recently deposited is partially swept away and its place taken by coarser sediments; and (2) oblique lamination, or current-bedding, where the strata are horizontal as usual, but the component laminæ are inclined at various angles. This structure is characteristic of sediments swept along by strong currents, especially when deposited in shallow basins or depressions.

Fig. 7.—Double Unconformability: *q.* quartzite; *s.* sandstone; *d.* drift.

RIPPLE-MARKS.—All who have been on a beach or sand-bar must have noticed the lines of wavy ridges and hollows, or ripples, on the surface of the sand. These are sand-waves, produced by water moving over the sand, or by air moving over dry sand, as ordinary waves are formed by air moving over water. Each tide usually effaces the ripple-marks made by its predecessor and leaves a new series, to be obliterated by the next tide. But where sediment is constantly accumulating, a rippled surface may be gently overspread by a new layer, and thus preserved. Other series of ripples may, in like manner, be formed and preserved in overlying layers; and when the beach becomes a firm sandstone, a section of it will show the rippled surfaces almost as distinctly as when they were first formed (Fig. 8). Ripple-marks are most perfect in fine sand. They are not formed in gravel, because it is too coarse; nor in clay, because it is too tenacious. They are usually limited to shallow water; and are always regarded as proving that the rocks in which they occur are shallow-water or beach deposits. They are normally at right angles to the current that produces them, and where this changes with the direction of the wind, cross-ripples and other irregularities are introduced. Ripple-marks are also usually parallel with the beach, and when they are found in the rocks they give us the direction, as well as the position, of the ancient shore-line.

Again, the friction of the water pushes the sand-grains along, rolling them up on one side of the ripple and letting them fall down on the other. Hence ripples, formed by a current are always moving and are unsymmetrical on the cross-section, presenting a long, gentle slope toward the current, and a short, steep slope away from it, the arrow in the figure indicating the direction of the current, or of the sea in the case of a beach. And we may thus learn from the fossil ripples, in some cases, not only the position and direction of the ancient shore, but also on which side the land lay, and on which side the sea. When the water is in a state of oscillation, without any distinct current, more symmetrical ripples are produced.

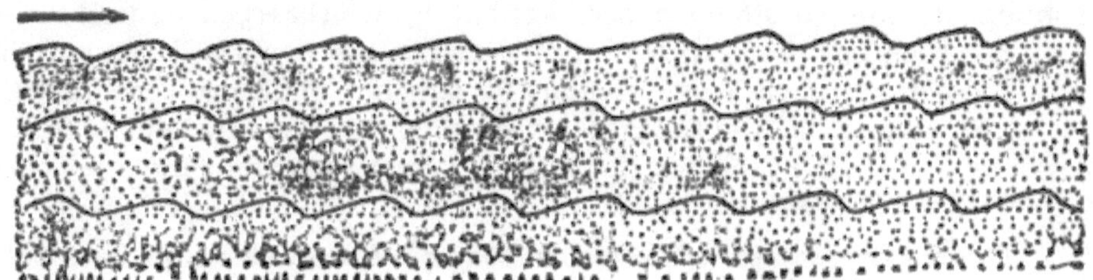

Fig. 8.—Ripple-marks in sandstone.

RILL-MARKS, RAIN-PRINTS, AND SUN-CRACKS.—"One of the most fascinating parts of the work of a field-geologist consists in tracing the shores of former seas and lakes, and thus reconstructing the geography of successive geological periods." His conclusions, as we have already seen, are based largely upon the nature of the sediments; but still more convincing is the evidence afforded by those superficial features of the strata, which, like ripple-marks, seem, by themselves, quite insignificant. And among these he lays special emphasis upon those which show that during their deposition strata have at intervals been laid bare to sun and air.

During ebb tide water which has been left at the upper edge of the beach runs down across the beach in small rills, which excavate miniature channels; and when these are preserved in the hard rocks, they prove that the latter are beach deposits, and, like the ripple-marks, show the direction of the old shore.

If a heavy shower of rain falls on a muddy beach or flat, the sediment deposited by the returning tide may cover, without obliterating, the small but characteristic impressions of the individual drops; and these markings are frequently found well preserved in the hardest slates and sandstones, testifying unequivocally to the conditions under which the rocks were formed. In some cases the rain-prints are found to be ridged up on one side only, in such a manner as to indicate that the drops as they fell were driven aslant by the wind. The prominent side of the marking, therefore, indicates the side towards which the wind blew.

Muddy sediments, especially in lakes and rivers, are often exposed to the air and sun during periods of drouth, and as they gradually dry up, polygonal cracks are formed. The sediment of the next layer will fill these sun-cracks; and when, as often happens, it is slightly different from the dessicated layer, they may still be traced. Sun-cracks preserved in this way are very characteristic of argillaceous rocks, and, of course, prove that in early times, as at the present day, sediments of this class were exposed by the temporary retreat of the water. The foot-prints or trails of land-animals are often, as in the sandstones and shales of the Connecticut Valley, associated with, and of course strongly corroborate, all these other evidences of shore deposits. From the foot-prints preserved in the rocks we pass naturally to the consideration of the fossil remains of plants and animals found entombed in the strata.

FOSSILS.—Although fossils find their highest interest in the light which they throw upon the succession of life on the globe, they may also be properly regarded as structural features of stratified rocks; and any one who has seen the dead shells, crabs, fishes, etc., on the beach will readily understand how fossils get into the rocks. It is not our province here to study the structure of the fossils themselves, for that would involve us in a course in paleontology, a task belonging to the biologist rather than the geologist; but we will merely observe the three principal degrees in the preservation of fossils:—

1. *Original composition not completely changed.*—Extinct elephants have been found frozen in the river-bluffs of Siberia so perfectly preserved that dogs and wolves ate their flesh. The bodies of animals are also found well preserved in peat-bogs. All coal is simply fossil vegetation retaining in a large degree the original composition; and the same is true of ferns, etc., preserved as black impressions in the rocks. All bones and shells consist of mineral matter which makes them hard, and animal matter which makes them tough and strong. In very many cases, especially in the newer formations, the animal matter is still partially, and the mineral matter almost wholly, intact.

2. *Original composition completely changed, but form and structure preserved.*—All kinds of fossils are commonly called petrifactions, but only those preserved in this second way are truly petrified, *i.e.*, turned to stone. "Petrified wood is the best illustration, and in a good specimen not only the external form of the wood, not only its general structure—bark, wood, radiating silver-grain, and concentric rings of growth—are discernible, but even the microscopic cellular structure of the wood, and the exquisite sculpturing of the cell-walls, are perfectly preserved, so that the kind of wood may often be determined by the microscope with the utmost certainty. Yet not one particle of the organic matter of the wood remains. It has been entirely replaced by mineral matter; usually by some form of silica. The same is true of the shells and bones of animals."—LE CONTE.

3. *Original composition and structure both obliterated, and form alone preserved.* — This occurs most commonly with shells, although fossil trees are also often good illustrations. The general result is accomplished in several ways: (*a*) The shell after being buried in the sediment may be removed by solution, leaving a *mould* of its external form, (*b*) This mould may subsequently be filled by the infiltration of finer sediment, forming a *cast* of the exterior of the shell. (*c*) The shell, before its solution, may have been filled with mud; and if the shell itself is then dissolved, we have a cast of its interior in a mould of its exterior.

TIME REQUIRED FOR THE FORMATION OF STRATIFIED ROCKS. — Many attempts have been made to determine the time required for the deposition of any given thickness of stratified rocks. Of course, only roughly approximate results can be hoped for in most cases; but these are at least sufficient to make it certain that geological time is very long. The average relative rate of growth of different kinds of sediment is, however, less open to doubt, for we have already seen that coarse sediments like gravel and sand accumulate much more rapidly than finer sediments like clay and limestone; and we are sometimes able to compare these two classes of rocks on a very large scale.

Thus, during what is known as the Paleozoic era, a sea extended from the Blue Ridge to the Rocky Mountains. Along the eastern margin of this sea, where the Alleghany Mountains now stand, sediments — chiefly conglomerate and sandstone, with some slate and less limestone — accumulated to a thickness of nearly 40,000 feet. Toward the west, away from the old shore-line, the coarse sediments gradually die out, and the formations become finer and thinner. In western Ohio and Indiana, slate and limestone predominate; while in the central part of the ancient sea, in Illinois and Missouri, the paleozoic sediments are almost wholly limestones, and have a thickness of only 4000 to 5000 feet. In other words, while one foot of limestone was forming in the Mississippi Valley, eight to ten feet of coarser sediments were deposited in Pennsylvania.

The best estimates show that coral-reefs rise — *i.e.*, limestones are formed on them — at the rate of about one foot in two hundred years. But coral limestones grow much more rapidly than limestones in general. Sandstones sometimes accumulate so rapidly that trees are buried before they have time to decay and fall (Fig. 9). Such a buried forest, like a coal-bed, represents a land surface, and proves a subsidence of the land; and in some cases, as indicated by the section, repeated oscillations of the crust may be proved in this way.

The mud deposited by the annual overflow of the Nile is forty feet thick near the ancient city of Memphis; and the pedestal of the statue of Rameses II., believed to have been erected B.C. 1361, is buried to a depth of nine feet, four inches, indicating that 13,500 years have elapsed since the Nile began to spread its mud over the sands of the desert.

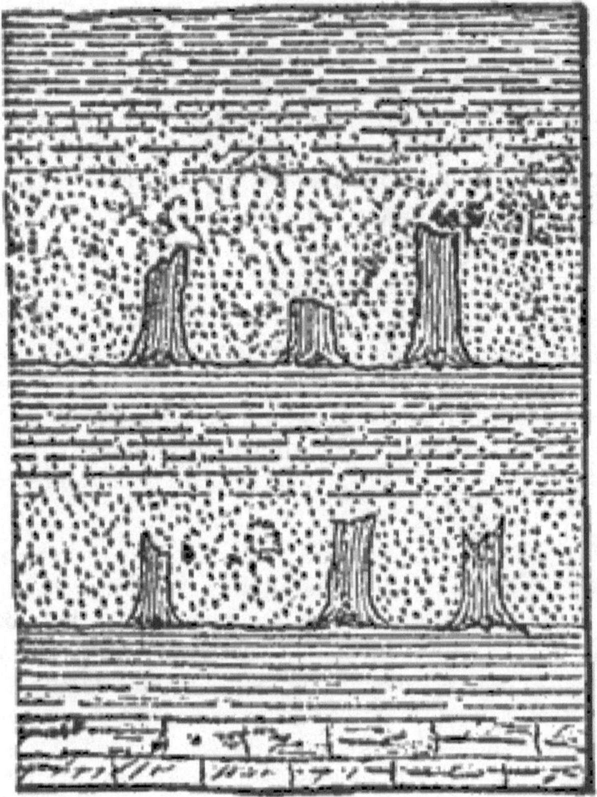

Fig. 9.—Erect fossil trees.

But the greatest difficulty in estimating the time required for the formation of any series of strata arises from the fact that we cannot usually even guess at the length of the periods when the deposition has been partially or wholly interrupted. Now and then, however, we find evidence that these periods may be very long. A layer of fossil shells in sandstone or slate proves an interruption of mechanical deposition. Beds of coal, fossil forests, and other indications of land surfaces are still more conclusive. The interposition of strata (Fig. 5) proves a prolonged interruption of deposition over the area not covered by the interposed bed. But the most important of all evidence is that afforded by unconformability; and the length of the lost interval between the two formations is measured approximately by the erosion of the older.

Original Structures of Eruptive Rocks.

The structures of this class are divisible into those pertaining to the volcanic rocks and those pertaining to the fissure or dike rocks. But since volcanoes are rare in this part of the world, while dikes are well developed in many sections of our country, it seems best to give our attention chiefly to the latter.

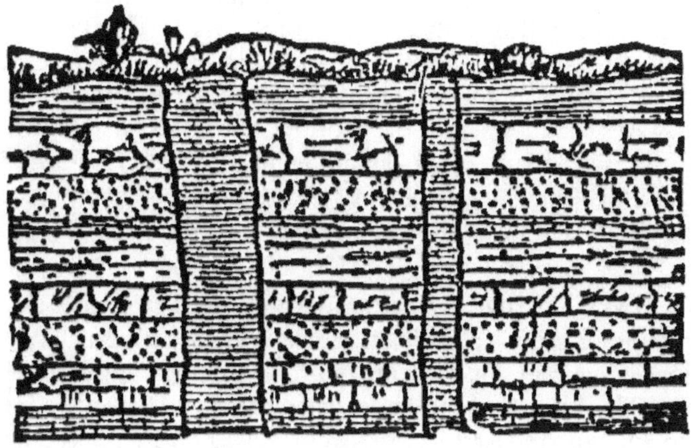

Fig. 10.—Typical dikes.

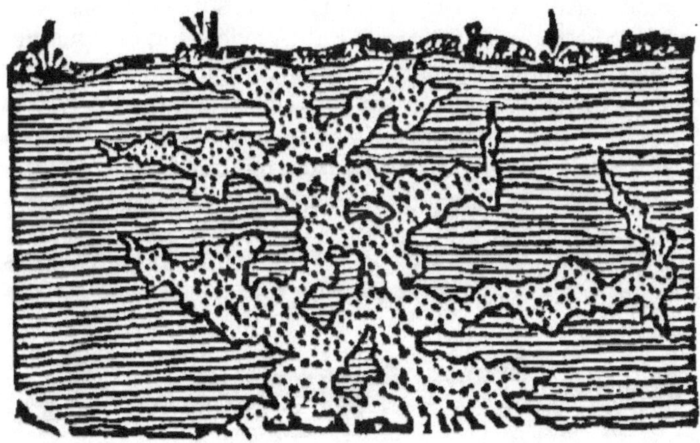

Fig. 11.—Section of a granite mass.

The term *dike* is a general name for all masses of eruptive rocks that have cooled and solidified in fissures or cavities in the earth's crust. But the name is commonly restricted to the more regular, wall-like masses (Fig. 10), those having extremely irregular outlines, like most masses of granite (Fig. 11), being known simply as *eruptive masses*. The propriety of this distinction is apparent when we consider the origin of *dike* as a geological term. It was first used in this sense in southern Scotland, where almost any kind of a wall or barrier is called a dike. The dikes traverse the different stratified formations like gigantic walls, which are often encountered by the coal-miners, and on the surface are frequently left in relief by the erosion of the softer enclosing rock, so that in the west of Scotland, especially, they are actually made use of for enclosures. In other cases the dike has decayed faster than the enclosing rock, and its position is marked by a ditch-like depression. The narrow, straight, and perpendicular clefts or chasms observed on many coasts are usually due to the removal of the wall-like dikes by the action of the waves. Dikes are sometimes mere plates of rock, traceable for a few yards only; and they range in size from that up to those a hundred feet or more in width, and traceable for scores of miles across the country, their outcrops forming prominent ridges. The sides of dikes are often as parallel and straight of those of built walls, the resemblance to human workmanship being heightened by the numerous joints which, intersecting each other along the face of a dike, remind us of well-fitted masonry.

FORMS OF DIKES.—A dike is essentially a casting. Melted rock is forced up from the heated interior into a cavity or crack in the earth's crust, cools and solidifies there, and, like a metallic casting, assumes the form of the fissure or mould. In other words, the form of the dike is exactly that of the fissure into which the lava was injected. Now the forms of fissures depend partly upon the nature of the force that produces them, but very largely upon the structure—and especially the joint-structure—of the enclosing rocks. Nearly all rocks are traversed by planes of division or cracks called joints, which usually run in several directions, dividing the rock into blocks. And it is probable that dike-fissures are most commonly produced, not by breaking the rocks anew, but by widening or opening the pre-existing joint-cracks. Hence the straight and regular jointing of slate, limestone and most sedimentary rocks is accompanied by wall-like dikes—the typical dikes (Fig. 10); while the more irregular jointing of granite and other massive rocks gives rise to sinuous, branching, variable dikes. The general dependence of dikes upon the joint-structure of the rocks is proved by the facts that dikes, like joints, are normally vertical or highly inclined, and that they are usually parallel with the principal systems of joints in the same district. The wall-like dikes also give off branches, but usually in a regular manner, as shown in Fig. 12.

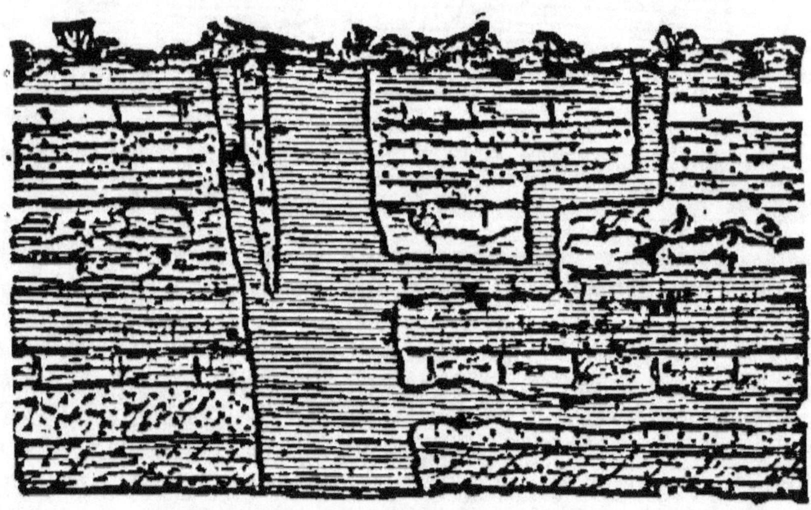

Fig. 12.—Dike with regular branches.

STRUCTURE OF DIKES.—The rock traversed by a dike is called the *country* or *wall* rock. Fragments of this are often torn off by the igneous material, and become enclosed in the latter. Such enclosed fragments may sometimes form the main part of the dike, which then, since they are necessarily angular, often assumes the aspect of a breccia. This is the only important exception to the rule that dikes are homogeneous in composition; *i.e.*, in the same dike we can usually find—from end to end, from side to side, and probably from top to bottom—no essential difference in composition. But there is often a marked contrast in *texture* between different parts of a dike, and especially between the sides and central portion. The liquid rock loses heat most rapidly where it is in contact with the cold walls of the fissure, and solidifies before it has time to crystallize, remaining compact and sometimes even glassy; while in the middle of the dike, unless it is very narrow, it cools so slowly as to develop a distinctly crystalline texture. There is no abrupt change in texture, but a gradual passage from the compact border to the coarsely crystalline or porphyritic middle portion. It is obvious that a similar gradation in texture must exist between the top and bottom of a dike.

CONTACT PHENOMENA.—Under this head are grouped the interesting and important phenomena observable along the contact between the dike and wall-rock. These throw light upon the conditions of formation of dikes, and are often depended upon to show whether a rock mass is a dike or not. The student will observe here:—

1. The detailed form of the contact. It may be straight and simple, or exceedingly irregular, the dike penetrating the wall, and enclosing fragments of it, as in Fig. 11, which is a typically igneous contact.

2. The alteration of the wall-rock by heat. This may consist in: (*a*) *coloration*, shales and sandstones being reddened in the same way as when clay is burnt for bricks; (*b*) *baking and induration*, sandstone being converted into quartzite and even jasper; clay, slate, etc., being not only baked to a flinty hardness, but actually vitrified, as in porcelainite; and bituminous coal being converted into natural coke or anthracite; and (*c*) *crystallization*, chalk, and other limestones being changed to marble, and crystals of pyrite, calcite, quartz, etc., being developed in slate, sandstone, and other rocks.

3. The alteration of the dike-rock by (*a*) more rapid cooling, and (*b*) the access of thermal waters.

The alteration of the wall-rock may extend only a few inches or many yards from the dike, gradually diminishing with the distance; and the cases are surprisingly numerous where there is no perceptible alteration; and, again, the alteration is usually mutual, the dike-rock being altered in texture, color, and composition.

INTRUSIVE BEDS.—We commonly think of dikes as cutting across the strata, but they often lie in planes parallel with them; and the same dike may run across the beds in some parts of its course and between them in others (Fig. 12), or the conformable dike maybe simply a lateral branch of a main vertical dike, as shown in the same figure. All dikes or portions of dikes lying conformably between the strata are called *intrusive beds* or *sheets*.

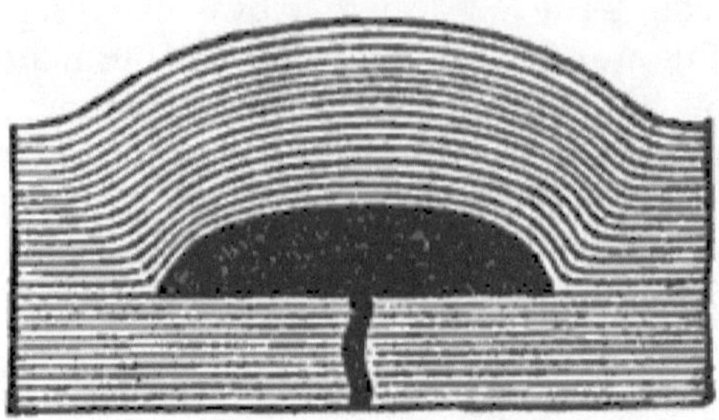

Fig. 13.—Ideal cross-section of a laccolite.

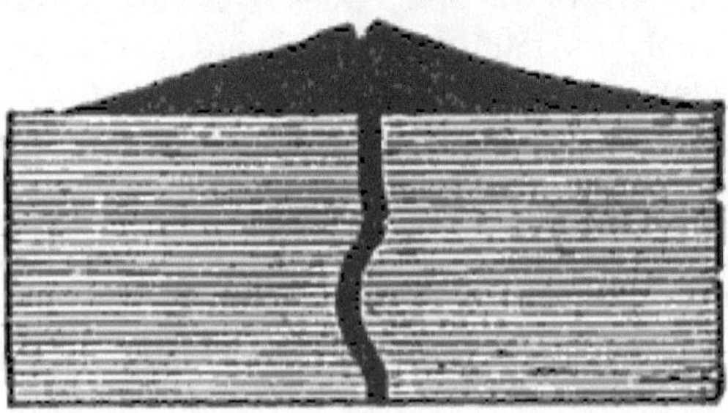

Fig. 14.—Ideal cross-section of a volcano.

When a dike fails to reach the surface, but spreads out horizontally between the strata, forming a thick dome or oven-shaped intrusive bed, the latter is called a *laccolite* (Fig. 13). Laccolites are sometimes of immense volume, containing several cubic miles of rock. Fig. 14 enables us to compare the laccolite with the volcano.

In the one case a large mound of eruptive material accumulates between the strata, the overlying beds being lifted into a dome; while in the other case the fissure or vent reaches the surface, and the mound of lava is built up on top of the ground.

Cotemporaneous Beds.—When the lava emitted by a crater is sufficiently liquid, it spreads out horizontally, forming a volcanic sheet or bed. If such an eruption is submarine, or the lava flow is subsequently covered by the sea, sedimentary deposits are formed over it; and beds of lava which thus come to lie conformably between sedimentary strata are known as *cotemporaneous sheets* or *beds*, because they belong, in order of time, in the position in which we find them, being, like any member of a stratified series, newer than the underlying and older than the overlying strata. Cotemporaneous lava-flows are sometimes repeated again and again in the same district, and thus important formations are built up of alternating igneous and aqueous deposits. Evidently, the student who would read correctly the record of igneous activity in the past must be able to distinguish intrusive and cotemporaneous beds. The principal points to be considered in making this distinction are: (1) The intrusive bed is essentially a dike, dense and more or less crystalline in texture, altering, and often enclosing fragments of, both the underlying and overlying strata, and frequently jogging across or penetrating the sediments. (2) The cotemporaneous bed, on the other hand, being essentially a lava-flow, is much less dense and crystalline, being usually distinctly scoriaceous or amygdaloidal, especially at the borders, and the underlying strata alone showing heat action, or occurring as enclosures in the lava; for the overlying strata are newer than the lava, and often consist largely, at the base, of water-worn fragments of the lava.

AGES OF DIKES. — The ages of dikes may be estimated in several ways. They are necessarily newer than any stratified formation which they intersect or of which they enclose fragments; but any formation crossing the top of a dike must usually be regarded as newer than the dike, especially if it contains water-worn fragments of the dike rock.

The relative ages of different dikes are determined by their relations to the stratified formations; and still more easily by their mutual intersections, on the principle that when two dikes cross each other, the intersecting must be newer than the intersected dike. It is sometimes possible, in this way, to prove several distinct periods of eruption in the same limited district. The textures of dikes also often afford reliable indications of their ages; for, as we have already seen, the upper part of a dike, cooling rapidly and under little pressure, must be less dense and crystalline than the deep-seated portion, which cools slowly and under great pressure.

Now, the lower, coarsely crystalline part of a dike can usually be exposed on the surface only as the result of enormous erosion; and erosion is a slow process, requiring vast periods of time. Hence, when we see a coarse-grained dike outcropping on the surface, we are justified in regarding it as very old, for all the fine-grained upper part has been gradually worn away by the action of the rain, frost, etc. Other things being equal, coarse-grained must be older than fine-grained dikes; and the texture of a dike is at once a measure of its age and of the amount of erosion which the region has suffered since it was formed.

ERUPTIVE MASSES. — In striking contrast with the more or less wall-like dikes are the highly irregular, and even ragged, outlines of the eruptive masses; and it is worth while to notice the probable cause of this contrast. The true dikes are formed, for the most part, of comparatively fine-grained rocks — the typical "traps"; while the eruptive masses consist chiefly of the coarse-grained or granitic varieties. Now we have just seen that the coarse-grained rocks have been formed at great depths in the earth's crust, while the fine-grained are comparatively superficial. But we have good reason for believing that the joint-structure, upon which the forms of dikes so largely depend, is not well developed at great depths, where the rocks are toughened, if not softened, by the high temperature. In other words, trap dikes are formed in the jointed formations, which break regularly; while the granitic masses are formed where the absence of joint-structure and a high temperature combine to cause extremely irregular rifts and cavities when the crust is broken.

VOLCANIC PIPES OR NECKS. — Every volcano and every lava-flow or volcanic sheet must be connected with the earth's interior by a channel or fissure, which becomes a dike when the lava ceases to flow. But the converse proposition is not true, for it is probable that many dikes did not originally reach the surface, but have been exposed by subsequent denudation. This is conspicuously the case with laccolites and other forms of intrusive sheets. Volcanic sheets or beds have probably often resulted from the overflow of the lava at all points of an extensive fissure or system of fissures; but the vent of the true volcano must be more circumscribed, an approximately circular opening in the earth's crust, although doubtless originating in a fissure or at the intersection of two or more fissures, the lava continuing to flow at the widest part of the wound in the crust long after it has congealed in the narrower parts. Such a tube is known as the neck or pipe of the volcano; and volcanic necks are a highly interesting class of dikes, since they determine the exact location of many an ancient volcano, where the volcanic pile itself has long since been swept away. Necks and dikes are the downward prolongations or roots of the volcanic cone or sheet, and cannot be exposed on the surface until the volcanic fires have gone out and the agents of erosion have removed the greater part of the ejected materials.

Hence, equally with the dikes which originally failed to reach the surface, they, wherever open to our observation, testify to extensive erosion and a vast antiquity.

Original Structures of Vein Rocks.

Many things called veins are improperly so called, such as dikes of granite and trap, and beds of coal and iron-ore. The smaller, more irregular, branching dikes, especially, are very commonly called veins, and to distinguish the true veins from these eruptive masses, they are designated as *mineral veins* or *lodes*, although the term *lode* is usually restricted to the metalliferous veins.

ORIGIN OF VEINS. — Various theories of the formation of veins have been proposed, but the most of these are of historic interest merely, for geologists are now well agreed that nearly all true veins have been formed by the deposition of minerals from solution in fissures or cavities in the earth's crust. In many cases, especially where the veins are of limited extent, it seems probable that a part or all of the mineral matter was derived from the immediately enclosing rocks, being dissolved out by percolating water; and these are known as segregation or lateral secretion veins. But it is quite certain that as a general rule the mineral solutions have come chiefly from below, the deep-seated thermal waters welling up through any channel opened to them, and gradually depositing the dissolved minerals on the walls of the fissure as the temperature and pressure are diminished. This case, however, differs from the first only in deriving the vein-forming minerals from more remote and deeper portions of the enclosing rocks; and thus we see that vein-formation, whether on a large or a small scale, is always essentially a process of segregation.

We know that every volcano and every lava flow must be connected below the surface with a dike; and it is almost equally certain that the waters of mineral springs forming tufaceous mineral deposits on the surface, as in the geyser districts, also deposit a portion of the dissolved minerals on the walls of the subterranean channels, which are thus being gradually filled up and converted into mineral veins, which will be exposed on the surface when erosion has removed the tufaceous overflow. This connection of vein-formation with the superficial deposits of existing springs has been clearly proved in several important instances in Nevada and California.

Veins occur chiefly in old, metamorphic, and highly disturbed formations, where there is abundant evidence of the former existence of profound fissures, and in regions similar to those in which thermal springs occur to-day.

In the supplement to the lithological section the student will find the formation of a typical vein briefly described and contrasted with that of a typical dike; also a brief account of the lithological peculiarities of vein rocks, and general statements concerning their relative abundance and vast economic importance.

EXTERNAL CHARACTERISTICS OF VEINS.—The typical vein may be described as a fissure of indefinite length and depth, filled with mineral substances deposited from solution. Externally, it is very similar to the typical dike, for the fissures are made in the same way for both. Veins are normally highly inclined to the horizon; they exhibit in nearly every respect the same general relations to the structure of the country rock as dikes; and the ages of veins are determined in the same way as the ages of dikes.

The extensive mining operations to which veins have been subjected in all parts of the world, have made our knowledge of their forms below the surface very full and accurate. It has been learned in this way that very often the corresponding portions of the walls of a vein do not coincide in position, but one side is higher or lower than the other, showing that the walls slipped over each other when the fissure was formed or subsequently; and this faulting or displacement of the walls appears to be much more common with veins than with dikes, perhaps because the fissures remained open much longer. This slipping of the walls is the principal cause of the almost constant changes in the width of veins. For, since the walls are never true planes, and are often highly irregular any unequal movements must bring them nearer together at some points than at others. As a rule, the enormous friction accompanying the faulting, either crushes the wall-rock, or polishes and striates it, producing the highly characteristic surfaces known as *slicken-sides*. Where the wall is finely pulverized in this way, or is partially decomposed before or after the filling of the fissure, a thin layer of soft, argillaceous material is formed, separating the vein proper from the wall-rock. The miners call this the *selvage*; and it is a very characteristic feature of the true fissure veins.

Fragments of the wall-rock are frequently enclosed in veins, and the latter sometimes branch or divide in such a way as to surround a large mass of the wall, which is known as a "horse." A similar result is accomplished when a fissure is re-opened after being filled, if the new fissure does not coincide exactly with the old. It has been proved that veins have thus been re-opened and filled several times in succession; and in this way fragments of the older vein material become enclosed in the newer.

Although usually determined in direction by the joint-structure of the country rock, veins are often parallel with the bedding, especially in highly inclined, schistose formations. Such interbedded veins are commonly distinctly lenticular in form, occupying rifts in the strata which thin out in all directions and are often very limited in extent.

Whether conforming with the joint-structure or bedding, veins are commonly arranged in systems by their parallelism, those of different systems or directions usually differing in age and composition, and the older veins being generally faulted or displaced when intersected by the newer.

INTERNAL CHARACTERISTICS OF VEINS.—Internally, veins and dikes are strongly contrasted; and it is upon the internal features, chiefly, as previously explained, that we must depend for their distinction. In metalliferous veins the minerals containing the metal sought for (the galenite, sphalerite, etc.) are the *ore*; while the non-metalliferous minerals (the quartz, feldspar, calcite, etc.) are called the *gangue* or vein-stone proper. Although the combinations of minerals in veins are almost endless, yet certain associations of ores with each other and with different gangue minerals are tolerably constant, and constitute an important subject for the student of metallurgy and mining.

When a vein is composed of a single mineral, as quartz, it may rival a dike in its homogeneity. Most important veins, however, are composed of several or a large number of minerals, which may be sometimes more or less uniformly mixed with each other, but are usually distributed in the fissure in a very irregular manner. The great granite veins which are worked for mica, feldspar and quartz, are good illustrations, on a large scale, of the structure of veins in which several minerals have been deposited cotemporaneously. The individual minerals are found to a large extent, in great, irregular masses, with no order observable in their arrangement.

When a mineral is deposited from solution, it crystallizes by preference on a surface of similar composition, thus quartz on quartz, feldspar on feldspar, and so on; and it seems probable that this selective action of the wall-rock may be a principal cause of the irregular distribution of minerals in veins. It has often been observed in metalliferous veins that the richness varies with the nature of the adjacent country rock. This dependence of the contents of a fissure upon the wall-rock may be due in part to the selective deposition of the minerals, and in part to their derivation from the contiguous portions of the country or wall-rock, as in the so-called segregated veins. Temperature and pressure exert an important influence upon chemical precipitation, and it is, therefore, probable that the composition of many veins varies with the depth.

Frequently, perhaps usually, the minerals of composite veins are deposited in succession, instead of cotemporaneously, giving rise to the remarkable banded structure so characteristic of this class of veins. The first mineral deposited in the fissure forms a layer covering each wall, and is in turn covered by layers of the second mineral, and that by the third, and so on, until the fissure is filled, or the solution exhausted. The distinguishing features of this structure are shown in Fig. 15, in which *w w* represents the wall-rock, *a a*, *b b*, *c c* are successive layers of quartz, fluorite and barite, and the central band, *d*, is galenite. Since the vein grows from the outside inward, the outer layers are the oldest, and the central layers are the newest; again, the layers are symmetrically arranged, being repeated in the reverse order on opposite sides of the middle of the vein; and, lastly, in layers composed of prismatic crystals, as quartz (see the figure); the crystals are perpendicular to the wall and often project into, and even through, the succeeding layers. Such a crystalline layer is called a "*comb*" and the interlocking of the layers in this way is described as the *comb-structure* of the vein. The banding of veins is thus strongly contrasted with stratification, and with the structure in dikes due to the more rapid cooling along the walls. The duplicate layers are often discontinuous and of unequal thickness, on account of the strong tendency to segregation in the materials. This is clearly shown in Fig. 16, drawn on a reduced scale from a polished section of a lead vein in Cumberland, England, contained in the Museum of the Boston Society of Natural History. In this the gangue minerals are fluorite (f) and barite (b). The central band ($f\,g$) is a darker fluorite containing irregular masses of galenite. The banded structure of veins is exactly reproduced in miniature in the banding of agates, geodes, and the amygdules formed in old lavas. Unfilled cavities frequently remain along the middle of the vein. When small, these are known as "pockets." They are commonly lined with crystals; and when the latter are minute, the pockets are called druses. In metalliferous veins, the ore is much more abundant in some parts than in others, and these ore-bodies, especially when somewhat definite in outline, are known in their different forms and in different localities, as *courses*, *slants*, *shoots*, *chimneys*, and *bonanzas* of ore. The intersections and junctions of veins are often among the richest parts, as if the meeting of dissimilar solutions had determined the precipitation of the ore.

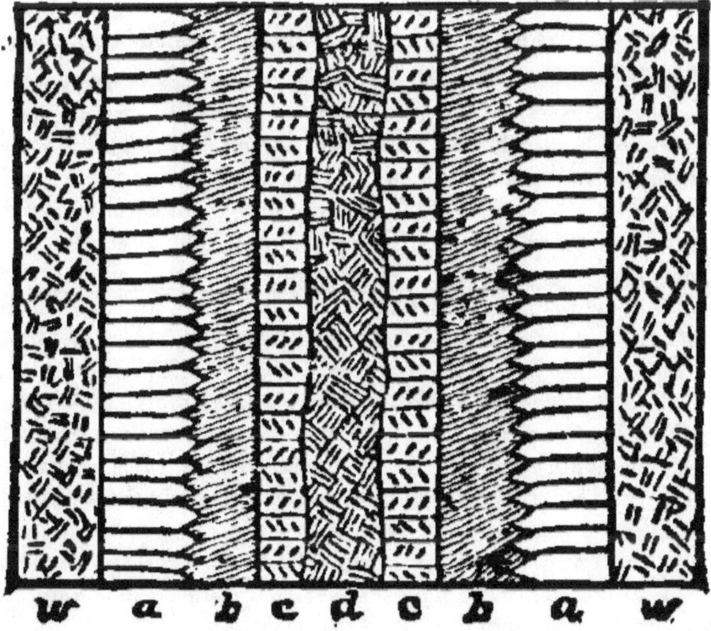

Fig. 15.—Ideal section of a vein.

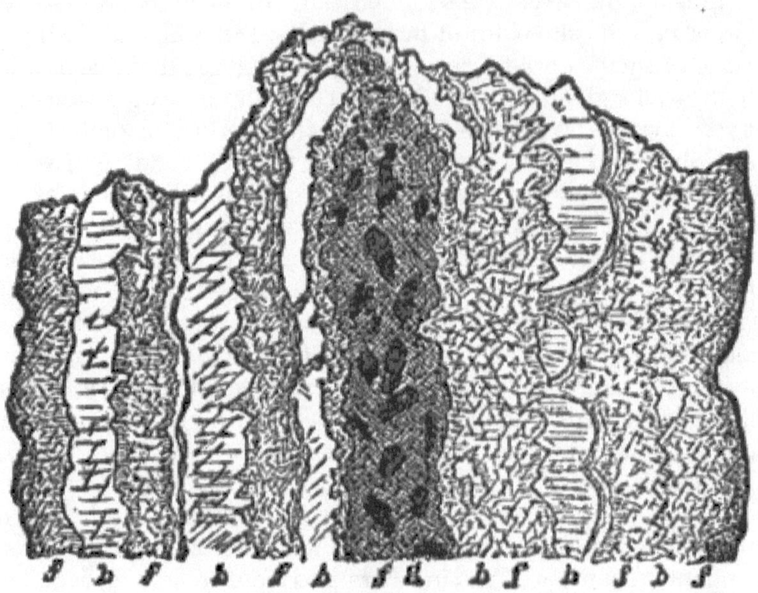

Fig. 16.—Section of a lead vein, one-fifth natural size.

Metalliferous veins, especially, are usually deeply decomposed along the outcrop by the action of atmospheric agencies. The ore is oxidized, and to a large extent removed by solution, leaving the quartz and other gangue minerals in a porous state, stained by oxides of iron, copper, and other metals, forming the *gossan* or *blossom-rock* of the vein.

Peculiar Types of Veins.—In calcareous or limestone formations, especially, the joint-cracks and bedding-cracks are often widened through the solution of the rock by infiltrating water, and thus become the channels of a more or less extensive subterranean drainage, by which they are more rapidly enlarged to a system of galleries and chambers, and, in some cases, large limestone caverns. The water dripping into the cavern from the overlying limestone is highly charged with carbonate of lime, which is largely deposited on the ceiling and floor of the cavern, forming stalactitic and stalagmitic deposits. These are masses of mineral matter deposited from solution in cavities in the earth's crust, and are essentially vein-formations. Portions of caverns deserted by the flowing streams by which they were excavated, are often filled up in this way, being converted into irregular veins of calcite. But calcite is not the only mineral found in these cavern deposits, for barite and fluorite, and various lead and zinc ores, especially the sulphides of these metals—galenite and sphalerite—have also been leached out of the surrounding limestone and concentrated in the caverns. The celebrated lead mines of the Mississippi Valley, and some of the richest silver-lead mines of Utah and Nevada are of this character. The forms of these cavern-deposits vary almost indefinitely, and are often highly irregular. The principal types are known as *gash-veins*, *flats* and *sheets* (Fig. 17), *chambers* and *pockets*.

Where joints and other cracks have opened slightly in different directions and become filled with infiltrated ores, we have what the German miners call a *stock-work*,—an irregular network of small and interlacing veins.

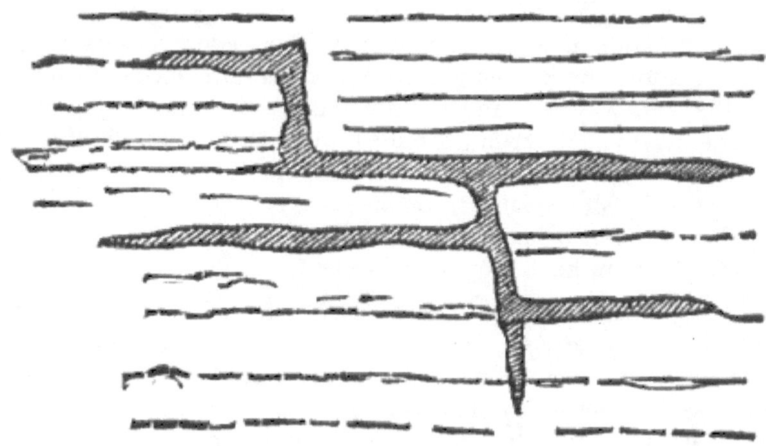

Fig. 17.—Gash-veins and sheets.

An *impregnation* is an irregular segregation of metalliferous minerals in the mass of some eruptive or crystalline rock. Its outlines are not sharply defined, but it shades off gradually into the enclosing rock.

Fahlbands are similar ill-defined deposits or segregations in stratified rocks. An impregnation or vein occurring along the contact between two dissimilar rocks is called a *contact deposit*. These are usually found between formations of different geological ages, and especially between eruptive and sedimentary rocks.

Subsequent Structures produced by Subterranean Agencies.

The subterranean forces concerned in the formation of rocks are chiefly various manifestations of that enormous tangential pressure developed in the earth's crust, partly by the cooling and shrinkage of its interior, but largely, it is probable, by the diminution of the velocity of the earth's rotation by tidal friction, and the consequent diminution of the oblateness of its form. It is well known that the centrifugal force arising from the earth's rotation is sufficient to change the otherwise spherical form of the earth to an oblate spheroid, with a difference of twenty-six miles between the equatorial and polar diameters. It is also well known that while the earth turns from west to east on its axis, the tidal wave moves around the globe from east to west, thus acting like a powerful friction-brake to stop the earth's rotation. Our day is consequently lengthening, and the earth's form as gradually approaching the perfect sphere. This means a very decided shortening and consequent crumpling of the equatorial circumference, and is equivalent to a marked shrinkage of the earth's interior, so far as the equatorial regions are concerned.

The most important and direct result of the horizontal thrust, whether due to cooling or tidal friction, is the corrugation or wrinkling of the crust; and the earth-wrinkles are of three orders of magnitude,—continents, mountain-ranges, and rock-folds or arches.

Continents and ocean-basins, although the most important and permanent structural features of the earth's crust, do not demand further consideration here, since their forms and relations are adequately described in the better text-books of physical geography. The forms and distribution of mountain-ranges might be dismissed in the same way; but, unlike continents, the structure of mountains, upon which their reliefs mainly depend, is quite fully exposed to our observation, and is one of the most important fields of the student of structural geology. Mountains, however, as previously explained, combine nearly all the kinds of structure produced by the subterranean agencies, and their consideration, therefore, belongs at the end rather than the beginning of this section.

Inclined or Folded Strata.—Normally, strata are horizontal, and dikes and veins are vertical or nearly so. Hence the stratified rocks are more exposed to the crumpling action of the tangential pressure in the earth's crust than the eruptive and vein rocks; and it is for this reason and partly because the stratified rocks are vastly more abundant than the other kinds, that the effects of the corrugation of the crust are studied chiefly in the former. But it should be understood that folded dikes and veins are not uncommon.

That the stratified rocks have, in many instances, suffered great disturbance subsequent to their deposition, is very evident; for, while the strata must have been originally approximately straight and horizontal, they are now often curved, or sharply bent and contorted, and highly inclined or even vertical. All inclined beds or strata are portions of great folds or arches. Thus we may feel sure when we see a stratum sloping downward into the ground, that its inclination or dip does not continue at the same angle, but that at some moderate depth it gradually changes and the bed rises to the surface again. Similarly, if we look in the opposite direction and think of the bed as sloping upward—we know that the surface of the ground is being constantly lowered by erosion, and consequently that the inclined stratum formerly extended higher than it does now, but not indefinitely higher; for, in imagination, we see it curving and descending to the level of the present surface again. Hence it forms, at the same time, part of one side of a great concave arch, and of a great convex arch, just as every inclined surface on the ground indicates both a hill and a valley. And guided by this principle we can often reconstruct with much probability folds that have been more or less completely swept away by erosion, or that are buried beyond our sight in the earth's crust.

The arches of the strata are rarely distinctly indicated in the topography, but must be studied where the ground has been partly dissected, as in cliffs, gorges, quarries, etc. They are also, as a rule, far more irregular and complex than they are usually conceived or represented. The wrinkles of our clothing are often better illustrations of rock-folds than the models and diagrams used for that purpose. This becomes self-evident when we reflect that the earth's crust is exceedingly heterogeneous in composition and structure, and must, therefore, yield unequally to the unequal strains imposed upon it.

The folds or undulations of the strata may be profitably compared with water-waves. In fact, the comparison is so close that they have been not inaptly called rock-waves. Folds, like waves, unless very large, rarely continue for any great distance, but die out and are replaced by others, giving rise to the *en echelon* or step-like arrangement. The plan of both a wave and a fold is a more or less elongated ellipse, each stratum in a fold being semi-ellipsoidal or boat-shaped. In other words, a normal fold is an elongated mound of concentric strata, being highest at the centre, sloping very gradually toward the ends, and much more abruptly toward the sides.

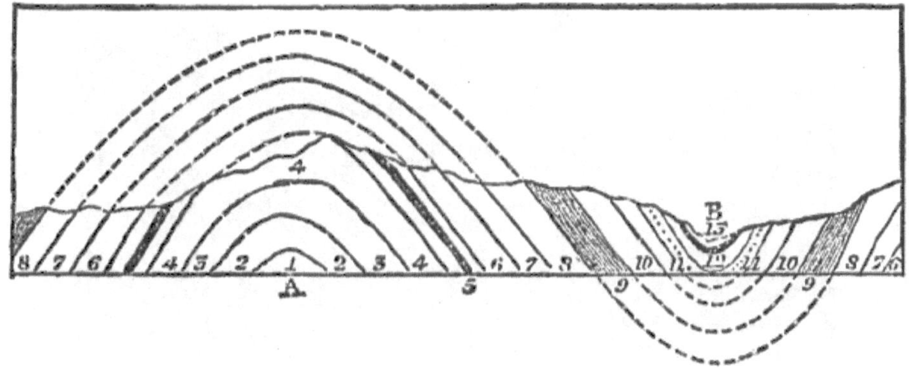

Fig. 18.—Anticlinal and synclinal folds.

The imaginary line passing longitudinally through a fold, about which the strata appear to be bent, is the *axis*; and the plane lying midway between the two sides of a fold and including the axis is the *axial plane*. The two principal kinds of folds are the *anticline* (Fig. 18, A), where the strata dip away from the axis; and the *syncline* (Fig. 18, B), where they dip toward the axis. They are commonly, but not always, correlative, like hill and valley.

Rock-folds are of all sizes, from almost microscopic wrinkles to great arches miles in length and breadth, and thousands of feet in height. The smaller folds, or such as may be seen in hand specimens and even in considerable blocks of stone, are commonly called contortions, and it is interesting to observe that they are, in nearly everything except size, precisely like the large folds, so that they answer admirably as geological models. Large folds, however, are almost necessarily curves, but contortions are frequently angular (Fig. 19). With folds, as with waves, the small undulations are borne upon the large ones; but the contortions are not uniformly distributed. An inspection of Fig. 18 shows that when the rocks are folded they must be in a state of tension on the anticlines (*A*), and in a state of compression in the synclines (*B*), and the latter is evidently the normal position of the puckerings or contortions of the strata, as shown in Fig. 20. Contortions are also most commonly found in thin-bedded, flexible rocks, such as shales and schists. And when we find them in hard, rigid rocks, like gneiss and limestone, it must mean either that the structure was developed with extreme slowness, or that the rock was more flexible then and possibly plastic.

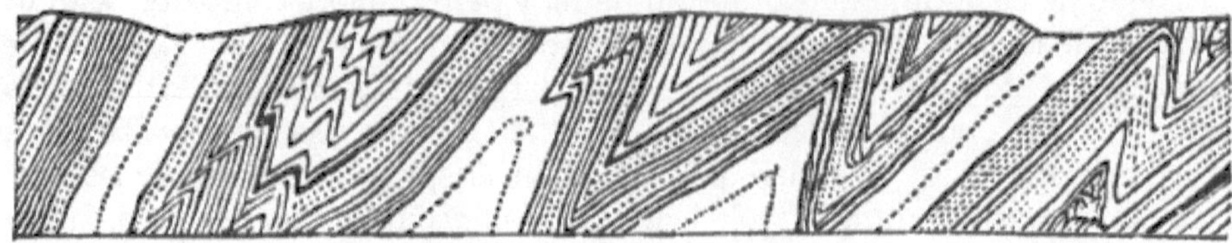

Fig. 19.—Contorted strata.

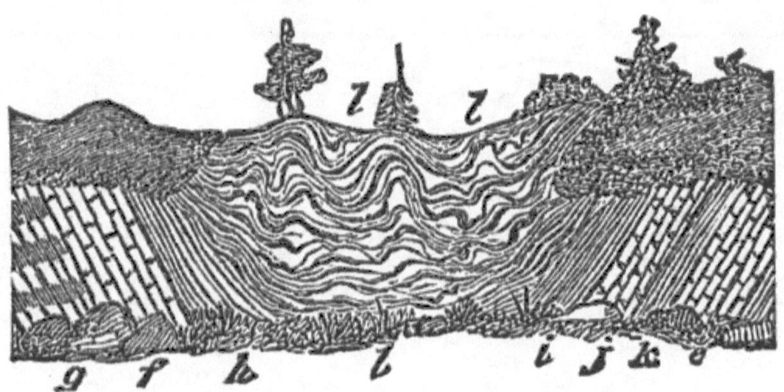

Fig. 20.—Contorted syncline.

Fig. 21.—Section of anticlinal mountains.

It is very interesting to notice the relations of anticlinal and synclinal folds to the agents of erosion. At the time the folds are made, the anticlinals, of course, are ridges, and the synclinals, valleys, and this relation sometimes continues, as shown in Fig. 21; but we have seen that the rocks in the trough of the synclinal are compressed and compacted, *i.e.*, made more capable of resisting erosion, while those on the crest of the anticlinal are stretched and broken, *i.e.*, made more susceptible of erosion. The consequence is that the anticlinals are usually worn away very much faster than the synclinals; so much faster that in many cases the topographic features are completely transposed, and in place of anticlinal ridges and synclinal valleys (Fig. 21) we find synclinal ridges and anticlinal valleys (Fig. 22).

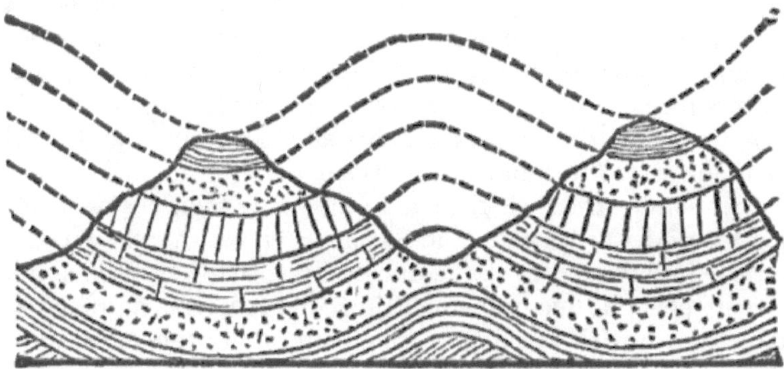

Fig. 22.—Section of synclinal mountains.

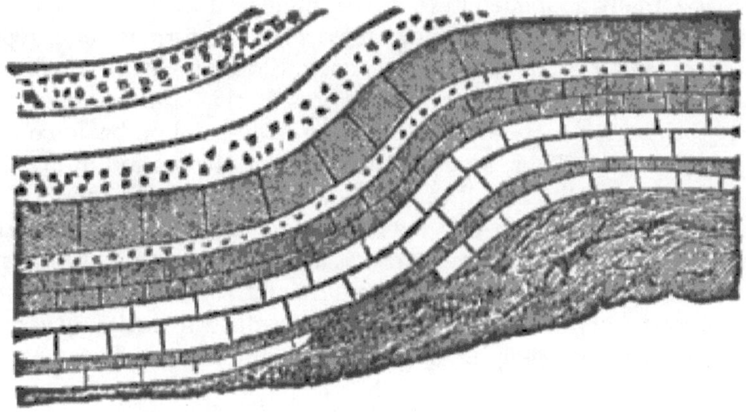

Fig. 23.—Monoclinal fold.

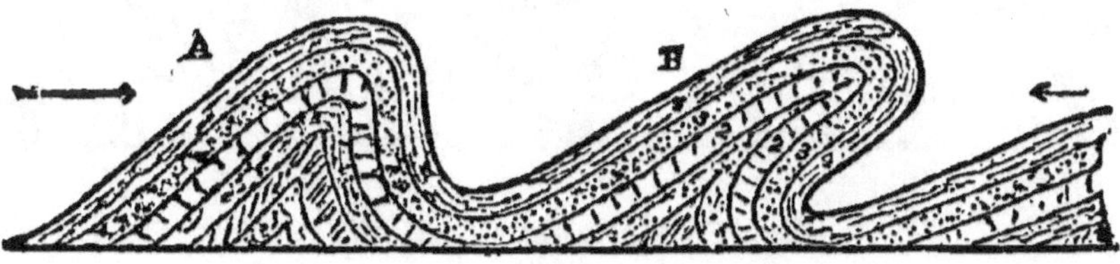

Fig. 24.—Unsymmetrical and inverted folds.

Besides the anticlinal and synclinal folds already explained, there are folds that slope in only one direction, one-sided or *monoclinal* folds (Fig. 23). Anticlinal and synclinal folds are *symmetrical* when the dip or slope of the strata is the same on both sides and the axial plane is vertical. The great majority of folds, however, are *unsymmetrical*, the opposite slopes being unequal, and the axial planes inclined to the vertical (Fig. 24, *A*). This means that the compressing or plicating force has been greater from one side than from the other, as indicated by the arrows. It acted with the greatest intensity on the side of the gentler slope, the tendency evidently having been to crowd or tip the fold over in the direction of the steep slope. When the steep slope approaches the vertical, this tendency is almost unresisted, and when it passes the vertical, gravitation assists in overturning the fold (Fig. 24, *B*). Such highly unsymmetrical folds, including all cases where the two sides of the fold slope in the same direction, are described as *overturned* or *inverted*, although the latter term is not strictly applicable to the entire fold, but only to the strata composing the under or lee side of it. Fig. 24, *B*, shows that these beds are completely inverted, the older, as the figures indicate, lying conformably upon the newer. This inversion is one of the most important features of folded strata, and it has led to many mistakes in determining their order of succession. In the great mountain-chains, especially, it is exhibited on the grandest scale, great groups of strata being folded over and over each other as we might fold carpets. An inverted stratum is like a flattened S or Z, and may be pierced by a vertical shaft three times, as has actually happened in some coal mines. Folds are *open* when the sides are not parallel, and *closed* when they are parallel, the former being represented by a half-open, and the latter by a closed, book. Closed folds are usually inverted, and when the tops have been removed by erosion (Fig. 25), the repetition of the strata may escape detection, and the thickness of the section be, in consequence, greatly overestimated. Thus, a geologist traversing the section in Fig. 25 would see thirty-two strata, all inclined to the left at the same angle, those on the right apparently passing below those on the left, and all forming part of one great fold. The repetition of the strata in reverse order, as indicated by the numbers, and the structure below the surface, show, however, that the section really consists of only four beds involved in a series of four closed folds, the true thickness of the beds in this section being only one-eighth as great as the apparent thickness.

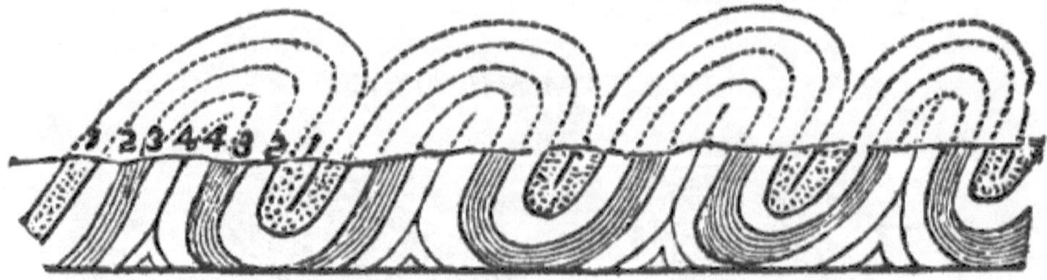

Fig. 25.—Series of closed folds.

The most important features to be noted in observing and describing inclined or folded strata are the *strike* and *dip*. The strike is the compass bearing or horizontal direction of the strata. It is the direction of the outcrop of the strata where the ground is level. It may also be defined as the direction of a level line on the surface of a stratum, and is usually parallel with the axis of the fold.

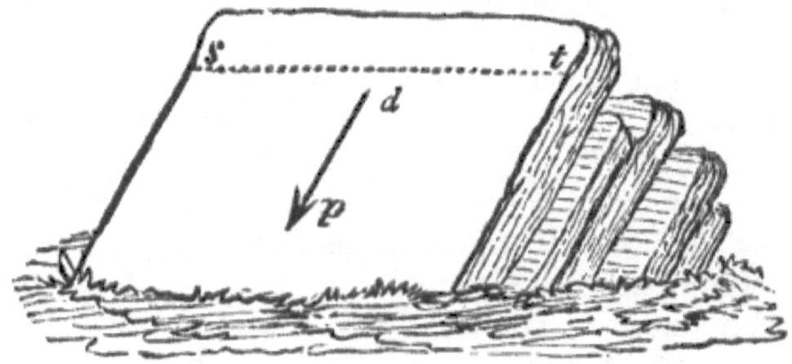

Fig. 26.—Dip and strike.

The dip is the inclination of the beds to the plane of the horizon, and embraces two elements: (*a*) the direction of the dip, which is always at right angles to the strike, being the line of steepest descent on the surface of the stratum, and (*b*) the amount of the dip, which is the value of the angle between the line of steepest descent and the horizon.

In Fig. 26, *s t* is the direction of the strike, and *d p* that of the dip. The strike and direction of the dip are determined with the compass, and the amount of the dip with the clinometer, an instrument for measuring vertical angles.

The strike is much less variable than the dip, being often essentially constant over extensive districts; while the dip, except in very large or closed folds, is constantly changing in direction and amount.

When the dip and surface breadth of a series of strata have been measured, it is a simple problem in trigonometry to determine the true thickness, and the depth below the surface of any particular stratum at any given distance from its outcrop. When the strata are vertical, the surface breadth or traverse measure is equal to the thickness.

By the *outcrop* of a stratum or formation we ordinarily understand its actual exposure on the surface, where it projects through the soil in ledges or quarries. But the term is also more broadly defined to mean the exposure of the stratum as it would appear if the soil were entirely removed. It is instructive to observe the relations of the outcrop to the form of the surface. Its breadth varies with its inclination to the surface, appearing narrow and showing its true thickness where it is perpendicular to the surface, and broadening out rapidly where the surface cuts it obliquely. The outcrops of horizontal strata form level lines or bands along the sides of hills and valleys, essentially contour lines in the topography; and appear as irregular, sinuous bands bordering the streams and valleys in the map-view of the country. The outcrops of vertical strata, dikes, or veins, on the other hand, are represented by straight lines and bands on the map. While the outcrops of inclined strata are deflected to the right or left in crossing ridges and valleys, according to the direction and amount of their inclination.

A geological map shows the surface distribution of the rocks, *i.e.*, gives in one view the forms and arrangement of the outcrops of all the rocks in the district

mapped, including the trend or strike of the folded strata. The map may be lithological, each kind of rock, as granite, sandstone, limestone, etc., being represented
by a different color; or, it may be historical, each color representing one geological
formation, *i.e.*, the rocks formed during one period of geological time, without reference to their lithological character. But in the best maps these two methods are
combined. The geological section shows the arrangement of the rocks below the
surface, revealing the dip of the strata and supplementing the map, both modes of
representation, the horizontal and vertical, being required to give a complete idea
of the geological structure of a country. For a detailed and satisfactory explanation
of the construction and use of geological maps and sections, students are referred
to Prof. Geikie's "Outlines of Field Geology."

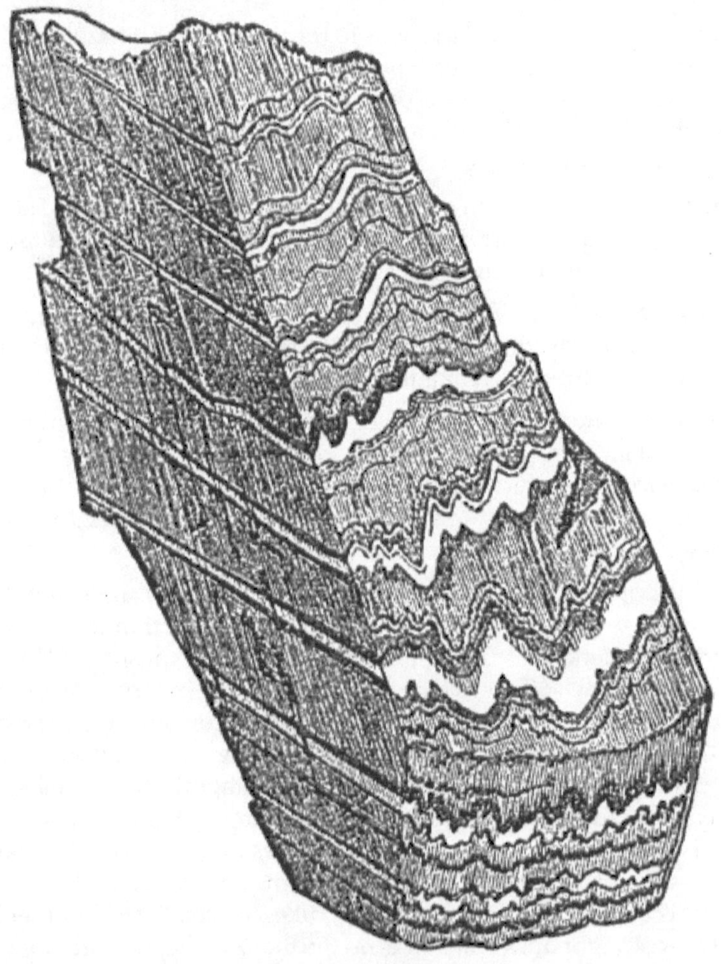

Fig. 27.—Slaty cleavage in contorted strata.

CLEAVAGE STRUCTURE.—This important structure is now known to be, like
rock-folds, a direct result of the great horizontal pressure in the earth's crust. It is

entirely distinct in its nature and origin from crystalline cleavage, and may properly be called lithologic cleavage. It is also essentially unlike stratification and joint-structure. It agrees with stratification in dividing the rocks into thin parallel layers, but the cleavage planes are normally vertical instead of horizontal. And the cleavage planes differ from joints in running in only one direction, dividing the rock into layers; while joints, as we shall see, traverse the same mass of rock in various directions, dividing it into blocks.

The principal characteristics of lithologic cleavage are: (1) It is rare, except in fine-grained, soft rocks, having its best development in the slates, roofing slates and school slates affording typical examples. Hence it is commonly known as *slaty cleavage*. (2) The cleavage planes are highly inclined or vertical, very constant in dip and strike, and quite independent of stratification. (3) It is usually associated with folded strata, and often with distorted nodules or fossils. The more important of these characteristics are illustrated by Fig. 27. This represents a block of contorted strata in which the dark layers are slate with very perfect cleavage parallel to the left-hand shaded side of the block; while the white layers are sandstone and quite destitute of cleavage. Many explanations of this interesting structure have been proposed, but that first advanced by Sharpe may be regarded as fully established. He said that *slaty cleavage is always due to powerful pressure at right angles to the planes of cleavage*. All the characteristics of cleavage noted above are in harmony with this theory. Cleavage is limited to fine-grained or soft rocks, because these alone can be modified internally by pressure, without rupture. Harder and more rigid rocks may be bent or broken, but they appear insusceptible of minute wrinkling or other change of structure affecting every particle of the mass. Since the cleavage planes are normally vertical, the pressure, according to the theory, must be horizontal. That this horizontal pressure exists and is adequate in direction and amount, is proved by the folds and contortions of the cleaved strata; for, as shown in Fig. 27, the cleavage planes coincide with the strike of the foldings, and are thus perpendicular to the pressure horizontally as well as vertically. The distortion of the fossils in cleaved slates is plainly due to pressure at right angles to the cleavage, for they are compressed or shortened in that direction, and extended or flattened out in the planes of cleavage. Again, Tyndall has shown that the magnetism of cleaved slate proves that it has been powerfully compressed perpendicularly to the cleavage. And, finally, repeated experiments by Sorby and others have proved that a very perfect cleavage may be developed in clay (unconsolidated slate) by compression, the planes of cleavage being at right angles to the line of pressure. When, however, Sharpe's theory had been thus fully demonstrated, the question as to *how* pressure produces cleavage still remained unanswered. Sorby held that clay contains foreign particles with unequal axes, such as mica-scales, etc., and that these are turned by the pressure so as to lie in parallel planes perpendicular to its line of action, thus producing easy splitting or cleavage in those planes. And he proved by experiments that a mixture of clay and mica-scales does behave in this way. But Tyndall showed that the cleavage is more perfect just in proportion as the clay is free from foreign particles, and in such a perfectly homogeneous substance as beeswax, he developed a more perfect cleavage than is possible in clay. His theory, which is now universally accepted, is, that the clay itself is composed of grains which are flattened by pressure, the granular structure with irregular fracture in all directions, changing to a scaly structure with very easy and plane fracture or splitting in one definite direction.

Observations on distorted fossils and nodules have shown that when slaty cleavage is developed, the rock is, on the average, reduced in the direction of the pressure to two-fifths of its original extent, and correspondingly extended in the vertical direction. Thus, whether rocks yield to the horizontal pressure in the earth's crust, by folding and corrugation, or by the flattening of their constituent particles, they are alike shortened horizontally and extended vertically; and it is impossible to overestimate the importance of these facts in the formation of mountains.

Faults or Displacements.—We may readily conceive that the forces which were adequate to elevate, corrugate, and even crush vast masses of solid rock were also sufficient to crack and break them; and since the fractures indicate that the strains have been applied unequally, it will be seen that unequal movements of the parts must often result. If this unequal movement takes place, *i.e.*, if the rocks on opposite sides of a fracture of the earth's crust do not move together, but slip over each other, a *fault* is produced. The two sides may move in opposite directions, or in the same direction but unequally, or one side may remain stationary while the other moves up or down. It is simply essential that the movement should be unequal in direction, or amount, or both; that there should be an actual slip, so that strata that were once continuous no longer correspond in position, but lie at different levels on opposite sides of the fracture. The vertical difference in movement is known as the *throw, slip*, or *displacement* of the fault. Fault-fractures rarely approach the horizontal direction, but are usually highly inclined or approximately vertical. When the fault is inclined, as in Fig. 28, the actual slipping in the plane of the fault exceeds the vertical throw, for the movement is then partly horizontal, the beds being pulled apart endwise. The inclination of faults, as of veins and dikes, should be measured from the vertical and called the *hade*. Faults are sometimes hundreds of miles in length; and the throw may vary from a fraction of an inch to thousands of feet.

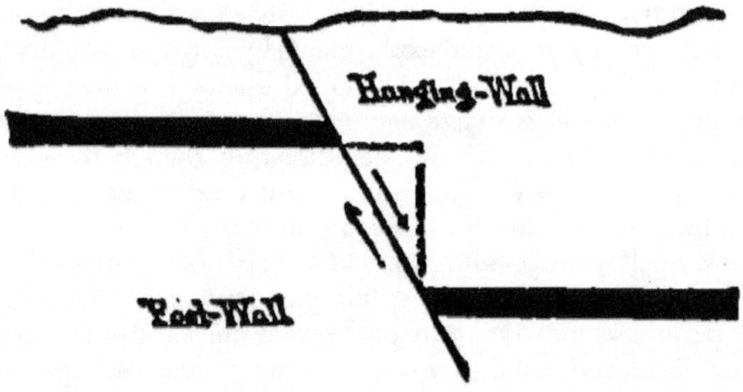

Fig. 28.—Section of a normal fault.

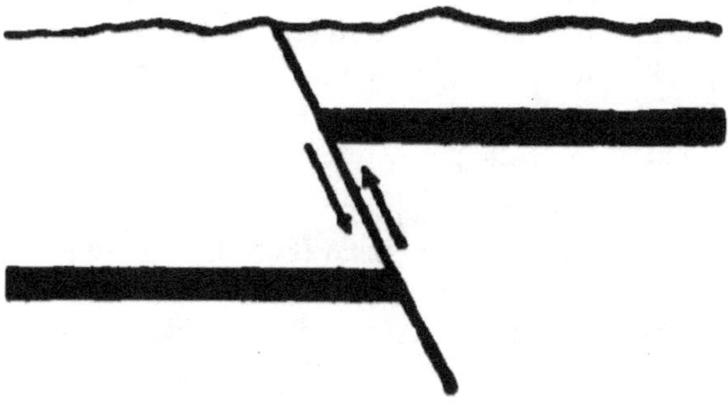

Fig. 29.—Section of a reversed fault.

Transverse sections, such as are represented by Fig. 28 and many specimens and models, do not give the complete plan or idea of a fault; but this is seen more perfectly in Fig. 30. We learn from this that a typical fault is a fracture along which the strata have *sagged* or settled down unequally. The most important point to be observed here is that the strata do not drop bodily, but are merely bent, the throw being greatest at the middle of the fault and gradually diminishing toward the ends. In other words, every simple fault must die out gradually; for we cannot conceive of a fault as ending abruptly, except where it turns upon itself so as to completely enclose a block of the strata, which may drop down bodily; but the fault is then really endless. A fault may be represented on a map by a line; if a simple fault, by a single straight line. But faults are often compound, and are represented by branching lines; that is, the earth's crust has been broken irregularly, and the parts adjoining the fracture have sagged or risen unequally.

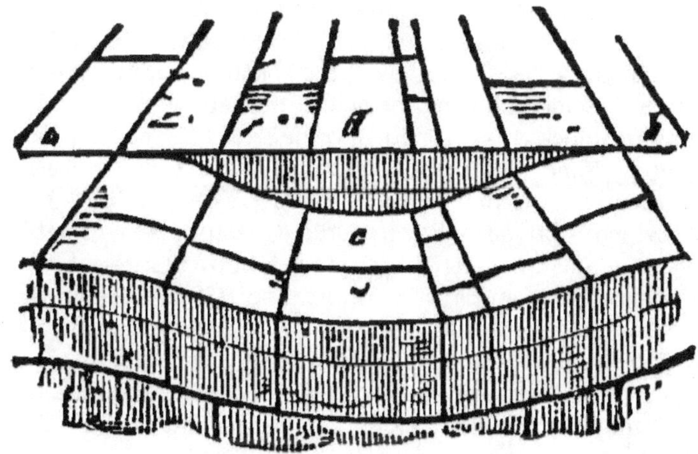

Fig. 30.—Ideal view of a complete fault.

The rock above an inclined fault, vein, or dike (Fig. 28) is called the *hanging wall*, and that below the *foot wall*. Now inclined faults are divided into two classes, according to the relative movements of the two walls. Usually, the hanging wall slips down and the foot wall slips up, as in Fig. 28. Faults on this plan are so nearly the universal rule that they are called *normal* faults. They indicate that the strata were in a state of tension, for their broken ends are pulled apart horizontally, so that a vertical line may cross the plane of a stratum without touching it.

A few important faults have been observed, however, in which the foot-wall has fallen and the hanging-wall has risen (Fig. 29). These are known as *reversed* faults; and they indicate that the strata were in a state of lateral compression, the broken ends of the beds having been pushed horizontally past each other, so that a vertical line or shaft may intersect the same bed twice, as has been actually demonstrated in the case of some beds of coal.

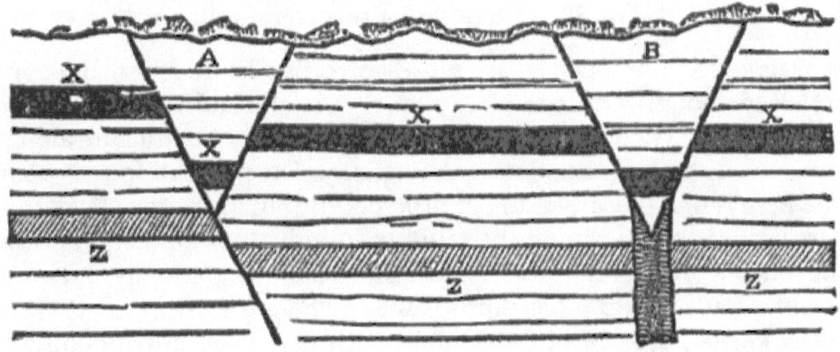

Fig. 31.—Explanation of normal faults.

The usual explanation of normal faults is given in Fig. 31. The inclined fractures of the earth's crust must often be converging, bounding, or enclosing large V-shaped blocks (*A, B*). If now, through any cause, as the folding of the strata, they are brought into a state of tension, so that the fractures are widened, the V-shaped masses, being unsupported, settle down, the fractures bounding them becoming normal faults, as is seen by tracing the bed X through the dislocations. The single fracture below the block *A* is inclined, and the stretching has been accomplished by slipping along it and faulting the bed Z as well as X, the entire section to the right of this fracture being part of a much larger V-shaped block the right-hand boundary of which is not seen. But the united fracture below the block *B* being vertical, any horizontal movement must widen it into a fissure, which is kept open by the great wedge above and may become the seat of a dike or mineral vein. The beds below the V may, in this case, escape dislocation, as is seen by tracing the bed Z across the fissure. These pairs of converging normal faults are called *trough* faults; and this is the only way in which we can conceive of important faults as terminating at moderate depths below the surface, and not affecting the entire thickness of the earth's crust.

Important reversed faults are believed to occur chiefly along the axes of overturned anticlines (Fig. 24) where the strata have been broken by the unequal strains, and those on the upper side shoved bodily over those on the lower or inverted side.

An extensive displacement of the strata is sometimes accomplished by short slips along each of a series of parallel fractures, producing a *step* fault.

Faults cutting inclined or folded strata are divided into two classes, according as they are approximately parallel with the direction of the dip or of the strike. The first are known as *transverse* or *dip* faults, and the second as *longitudinal* or *strike* faults. The chief interest of either class consists in their effect upon the outcrops of the faulted strata, after erosion has removed the escarpment produced by the dislocation.

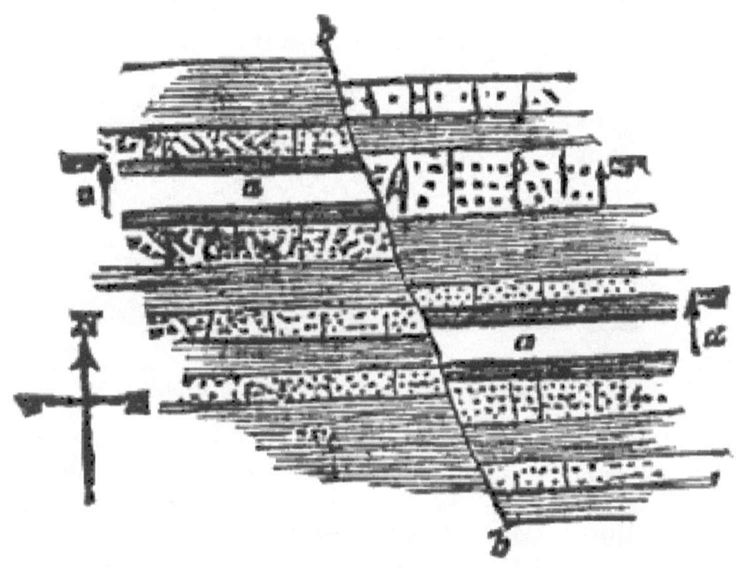

Fig. 32.—Plan of a dip fault.

Dip faults cause a lateral shift or displacement of the outcrops, as shown in Fig. 32, which represents a plan or map-view of the strata traversed by the fault *b b*, the down throw being on the right and the up throw on the left. The dip of the strata is indicated by the small arrows and the accompanying figures; and it will be observed on tracing the outcrop of any stratum, *a a*, across the fault that it is shifted to the right. If the throw of the fault were reversed, the displacement of the outcrop would be reversed, also. Strike faults are of two kinds, according as they incline in the same direction as the strata, or in the contrary direction. The effect of the first kind is to conceal some of the beds, as shown in Fig. 33, in which beds 5 and 6 do not outcrop, but we pass on the surface abruptly from 4 to 7. The apparent thickness of the section is thus less than the real thickness. When the fault inclines against the strata, on the other hand (Fig. 34), the outcrops of certain strata are repeated on the surface; and a number of parallel faults of this kind, a step fault, will, like a series of closed folds (Fig. 25), cause the apparent thickness of the section to greatly exceed the real thickness. Repetition of the strata by faulting is distinguished from repetition by folding by being in the same instead of the reverse order.

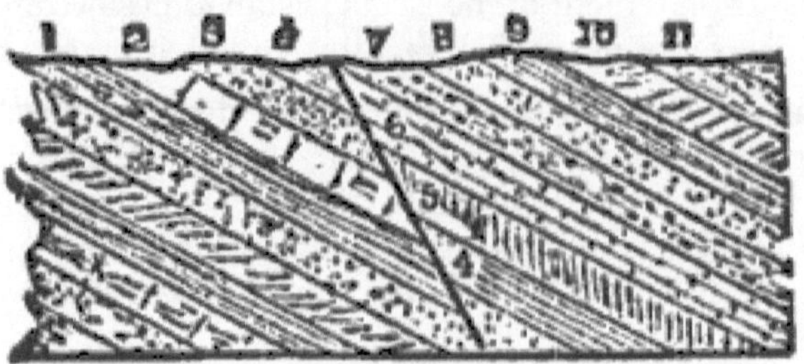

Fig. 33.—Strike fault, concealing strata.

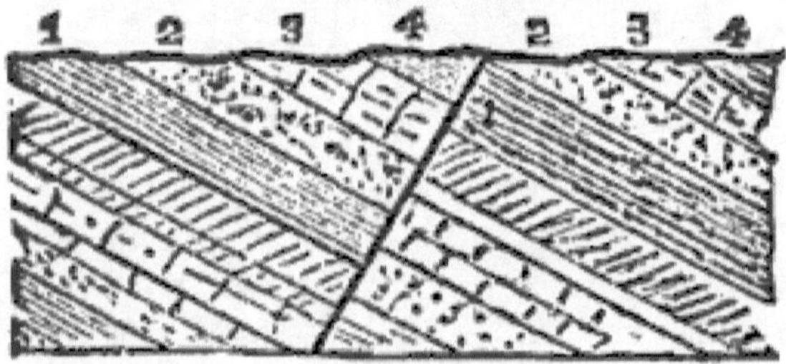

Fig. 34.—Strike fault, repeating strata.

Folds and faults are really closely related. In the former the strata are disturbed and displaced by bending; in the latter by breaking and slipping; and the displacement which is accomplished by a fold may gradually change to a fracture and slip. This relation is especially noticeable with monoclinal folds (Fig. 23), in which the tendency to shear or break the beds is often very marked.

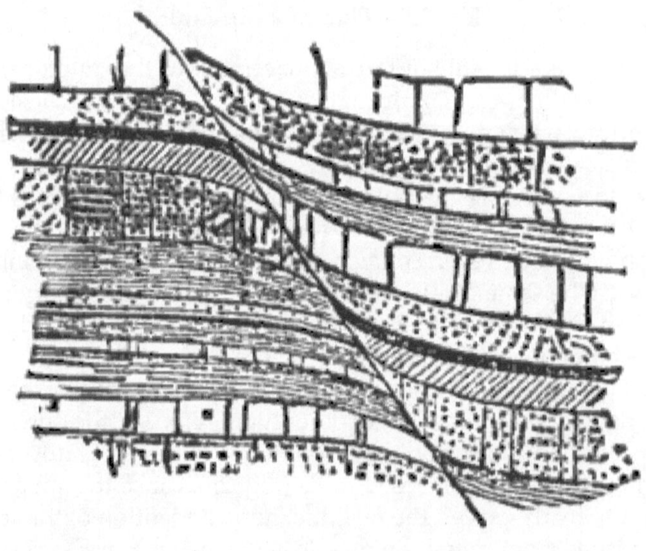

Fig. 35.—Section of beds distorted by a fault.

Important faults are rarely simple, well-defined fractures; but, in consequence of the enormous friction, the rocks are usually more or less broken and crushed, sometimes for a breadth of many feet or yards. The fragments of the various beds are then strung along the fault in the direction of the slipping, and this circumstance has been made use of in tracing the continuation of faulted beds of coal. In other cases the direction of the slip is plainly indicated by the bending of the broken ends of the strata (Fig. 35), and the beds are sometimes turned up at a high angle or even overturned in this way.

Since faults are not plane, but undulating and often highly irregular, fractures, the walls will not coincide after slipping; and if the rocks are hard enough to resist the enormous pressure, the cavities or fissures produced in this way may remain open. Now faults are continuous fractures of the earth's crust, reaching down to an unknown but very great depth; and hence they afford the best outlets for the heated subterranean waters; so that it is common to find an important fault marked on the surface by a line of springs, and these are often thermal. The warm mineral waters on their way to the surface deposit part of the dissolved minerals in the irregular fissures along the fault, which are thus changed to mineral veins. This agrees with the fact that the walls of veins usually show faulting as well as crushed rock, slickensides, and other evidences of slipping.

If the earth's surface were not subject to erosion, every fault would be marked on the surface by an escarpment equal in height to the throw of the fault; and, notwithstanding the powerful tendency of erosion to obliterate them, these escarpments are sometimes observed, although of diminished height. Thus, according to Gilbert, the Zandia Mountains in New Mexico are due to a fault of 11,000 feet, leaving an escarpment still 7000 feet high. But, as a rule, there is no escarpment or marked inequality of the surface, the fault, like the fold, not being distinctly indicated in the topography. In all such cases we must conclude either that the faults were made a very long time ago, or that they have been formed with extreme slowness, so slowly that erosion has kept pace with the displacement, the escarpments being worn away as fast as formed. These and other considerations make it quite certain that extensive displacements are not produced suddenly, but either grow by a slow, creeping motion, or by small slips many times repeated at long intervals of time.

Joints and Joint-structure. — This is the most universal of all rock-structures, since all hard rocks and many imperfectly consolidated kinds, like clay, are jointed. Joints are cracks or planes of division which are usually approximately vertical and traverse the same mass of rock in several different directions. They are distinguished from stratification planes by being rarely horizontal, and from both stratification and cleavage planes by being actual cracks or fractures, and by dividing the rock into blocks instead of sheets or layers. The art of quarrying consists in removing these natural blocks; and most of the broad flat surfaces of rock exposed in quarries, are the joint-planes (Fig. 36). Some of the most familiar features of rock-scenery are also due to this structure, cliffs, ravines, etc., being largely determined in form and direction by the principal systems of joints; and we have already seen that the same is true of veins and dikes.

Joints are divided by their characteristics and modes of origin into three classes as follows:—

Fig. 36.—Quarry showing two systems of parallel joints.

1. *The parallel and intersecting joints.*—This is by far the most important class, and has its best development in stratified rocks, such as sandstone, slate, limestone, etc. These joints are straight and continuous cracks which may often be traced for considerable distances on the surface. They usually run in several definite directions, being arranged in sets or systems by their parallelism. Thus in Fig. 36 one set of joints is represented by the broad, flat surfaces in light, and a second set crossing the first nearly at right angles, by the narrower faces in shadow. By the intersections of the different sets of joints the rock is divided into angular blocks.

Although many explanations of this class of joints have been proposed, it has long been the general opinion of geologists that they are due to the contraction of the rocks, *i.e.*, that they are shrinkage cracks. We shall soon see, however, that they lack the most important characters of cracks known to be due to shrinkage; and the present writer has advanced the view that movements of the earth's crust, and especially the swift, vibratory movements known as earthquakes, are a far more adequate and probable cause. It is well known that earthquakes break the rocks; and, if space permitted, it could be shown that the earthquake-fractures must possess all the essential features of parallel and intersecting joints.

2. *Contraction joints or shrinkage cracks.*—That many cracks in rocks are due to shrinkage, there can be no doubt. The shrinkage may result from the drying of sedimentary rocks; but more generally from the cooling of eruptive rocks. Every one has noticed in warm weather, the cracks in layers of mud or clay on the shore, or where pools of water have dried up; and we have already seen that these sun-cracks are often preserved in the hard rocks. They have certain characteristic features by which they may be distinguished from the joints of the first class.

They divide the clay into irregular, angular blocks, which often show a tendency to be hexagonal instead of quadrangular. The cracks are continually uniting and dividing, but are not parallel, and rarely cross each other. Sun-cracks never affect more than a few feet in thickness of clay, and are an insignificant structural feature of sedimentary rocks. In eruptive rocks, on the other hand, the contraction joints have a very extensive, and, in some cases, a very perfect development, culminating in the prismatic or columnar jointing of the basaltic rocks. This remarkable structure has long excited the interest of geologists, and, although the basalt columns were once regarded as crystals, and later as a species of concretionary structure, it is now generally recognized as the normal result of slow cooling in a homogeneous, brittle mass. The columns are normally hexagonal, and perpendicular to the cooling surface, being vertical in horizontal sheets and lava flows, as in the classic examples of the Giant's Causeway and Fingal's Cave, and horizontal in vertical dikes (Fig. 37). They begin to grow on the cooling surface of the mass and gradually extend toward the centre, so that dikes frequently show two independent sets of columns.

Fig. 37.—Columnar dike.

3. *The concentric joints of granitic rocks.*—In quarries of granite and other massive crystalline rocks, it is often very noticeable that the rock is divided into more or less regular layers by cracks which are approximately parallel with the surface of the ground, some of the granite hills having thus a structure resembling that

of an onion. The layers are thin near the surface, become thicker and less distinct downwards, and cannot usually be traced below a depth of fifty or sixty feet. These concentric cracks are of great assistance in quarrying, and are now regarded as due to the expansion of the superficial portions of the granite caused by the heat of the sun. In reference to this view of their origin these may be properly called *expansion joints*.

STRUCTURE OF MOUNTAIN-CHAINS.—Mountains are primarily of two kinds,—volcanic and non-volcanic. The structure of the former belongs properly with the original structures of the volcanic rocks; but the latter—the true mountains—owe their internal structure and altitude or relief almost wholly to the crumpling and mashing together of great zones of the earth's crust, being, as already pointed out, the culminating points of the plication, cleavage, and faulting of the strata. "A mountain-*chain* consists of a great plateau or bulge of the earth's surface, often hundreds of miles wide and thousands of miles long. This is usually more or less distinctly divided by great longitudinal valleys into parallel *ranges* and *ridges*; and these, again, are serrated along their crests, or divided into *peaks* by transverse valleys. In many cases this ideal chain is far from realized, but we have instead, a great bulging of the earth's crust composed on the surface of an inextricable tangle of ridges and valleys of erosion, running in all directions. In all cases, however, the erosion has been immense; for the mountain-chains are the great theatres of erosion as well as of igneous action. As a general fact, all that we see, when we stand on a mountain-chain—every peak and valley, every ridge and cañon, all that constitutes scenery—is wholly due to erosion."—LE CONTE.

The structure of mountains thus fells under two heads: (1) The internal structure and altitude, which are due to the action of the subterranean agencies. (2) The external forms, the actual relief, which are the product chiefly of the superficial agencies or erosion. The study of mountains has shown that: (1) They are composed of very thick sedimentary formations. Thus the sedimentary rocks have a thickness of 40,000 feet in the Alleghanies; of 50,000 feet in the Alps; and of two to ten miles in all important mountain-chains. Such thick deposits of sediments, as we have already seen, must be formed on a subsiding sea-floor, and in many mountain-chains, as in the Alleghanies, the great bulk of these sediments are still below the level of the sea. Again, thick sedimentary deposits can only be formed in the shallow, marginal portions of the sea; and when such a belt of thick shore deposits yields to the powerful horizontal thrust, and is crumpled and mashed up, it is greatly shortened in the direction of the pressure and thickened vertically, so that its upper surface is lifted high above the level of the sea, and a mountain-chain is formed and added to the edge of the continent. We thus find an explanation of the important fact that on the several continents, but notably on the two Americas, the principal mountain-ranges are near to and parallel with the coast lines.

2. The mountain-forming sediments are usually strongly folded and faulted, and exhibit slaty cleavage wherever they are susceptible of that structure; and the older rocks, especially, in mountains are often highly metamorphosed, and are traversed by numerous veins and dikes, the infallible signs of intense igneous activity.

"In other words, mountain regions have been the great theatres—(1) of sedimentation before the mountains were formed; (2) of plication and upheaval in the formation of the range; and (3) of erosion which determined the present outline. Add to these the metamorphism, the faults, veins, dikes, and volcanic outbursts, and it is seen that all geological agencies concentrate there."—Le Conte.

Since mountain-ranges are great up-swellings or bulgings of the strata, their structure is always essentially anticlinal; and they sometimes consist of a single more or less denuded anticline (Fig. 38), the oldest and lowest strata exposed forming the summit of the range. More commonly, however, the single great arch or uplift is modified by a series of longitudinal folds, as shown in the section of the Jura Mountains (Fig. 21). Still more commonly the folds are closely pressed together, overturned, broken, and almost inextricably complicated by smaller folds, contortions, and slips.

Fig. 38.—Anticlinal mountain.

The strata on the flanks of the mountains are usually less disturbed than those near the axis of the range, and are sometimes seen to rest unconformably against the latter. In this way it is proved that some ranges are formed by successive upheavals. But we have still more conclusive evidence that mountains are formed with extreme slowness in the fact that rivers sometimes cut directly through important ranges. This proves, first, that the river is older than the mountains; second, that the deepening of its channel has always kept pace with the elevation of the range.

Concretions and Concretionary Structure.—Folds, cleavage, faults, and joints—all the subsequent structures considered up to this point—are the product of mechanical forces. Chemical agencies, although very efficient in altering the composition and texture of rocks, are almost powerless as regards the development of rock-structures; and the only important structure having a chemical origin is that named above.

Concretions are formed by the segregation of one or more of the constituents of a rock. But there are three distinct kinds of segregation. If the water percolating through or pervading a rock, dissolves a certain mineral and afterwards deposits it in cavities or fissures, *amygdules*, *geodes*, or *veins* are the result. If the mineral is deposited about particular points in the mass of the rock, it may form *crystals*, the rock becoming *porphyritic*; or it may not crystallize, but build up instead the rounded forms called *concretions*, the texture or structure of the rock becoming *concretionary*. A great variety of minerals occur in the form of concretions, but this mode of occurrence is especially characteristic of certain constituents of rocks, such as calcite, siderite, limonite, hematite, and quartz. Concretions may be classified according to the nature of the segregating minerals; and in each class we may distinguish the *pure* from the *impure* concretions. A pure concretion is one entirely composed of the segregating mineral. Most nodules of flint and chert, quartz, geodes, concretions of pyrite, and many hollow iron-balls are good illustrations of this class. In all these cases the segregating mineral has been able in some way to remove the other constituents of the rock, and make room for itself. But in other cases it has lacked this power, and has been deposited between and around the grains of sand, clay, etc.; and the concretions are consequently *impure*, being composed partly of the segregating mineral, and partly of the other constituents of the rock. The calcareous concretions known as clay-stones are a good example of this class, being simply discs of clay, all the minute interstices of which have been filled with segregated calcite. The solid iron-balls are masses of sand filled in a similar manner with iron oxides.

Concretions are of all sizes, from those of microscopic smallness in some oölitic limestones up to those twenty-five feet or more in diameter in some sandstones.

The point of deposition, when a concretion begins to grow, is often determined by some concrete particle, as a grain or crystal of the same or a different mineral, a fragment of a shell, or a bit of vegetation, which thus becomes the nucleus of the concretion. The ideal or typical concretion is spherical; but the form is influenced largely by the structure of the rock. In porous rocks, like sandstone, they are frequently very perfect spheres; but in impervious rocks, like clay, they are flat or disc-shaped, because the water passes much more freely in the direction of the bedding than across it; while the concretions in limestones, the nodules of flint and chert, are often remarkable for the irregularity of their forms. In all sedimentary rocks the concretions are arranged more or less distinctly in layers parallel with the stratification, which usually passes undisturbed through the impure concretions. Many silicious and ferruginous concretions are hollow, apparently in consequence of the contraction of the substance after its segregation; and the shrinkage due to drying is still further indicated by the cracks in the septaria stones. The hollow, silicious concretions are usually lined with crystals (geodes), while the hollow iron-balls frequently enclose a smaller concretion. Rocks often have a concretionary structure when there are no distinct or separable concretions. And the appearance of a concretionary structure (pseudo-concretions) is often the result of the concentric decomposition of the rocks by weathering, as explained on page 13.

SUBSEQUENT STRUCTURES PRODUCED BY THE SUPERFICIAL OR AQUEOUS AGENCIES.—The superficial agencies, as we have seen in the section on dynamical geology, are, in general terms, water, air, and organic matter. Geologically considered, the results which they accomplish, may be summed up under the two heads of deposition and erosion—the formation of new rocks in the sea, and the destruction of old rocks on the land. In the rôle of rock-makers they produce the very important original structures of the stratified rocks; while as agents of erosion they develop the most salient of the subsequent structures of the earth's crust—the infinitely varied relief of its surface. As a general rule, to which recent volcanoes are one important exception, the original and subterranean structures of rocks are only indirectly, and often very slightly, represented in the topography; for this, as we have seen, is almost wholly the product of erosion. Therefore, what we have chiefly to consider in this section is to what extent and how erosion is influenced by the pre-existing structures of rocks.

Horizontal or very slightly undulating strata, especially if the upper beds are harder than those below, give rise by erosion to flat-topped ridges or table-mountains (Fig. 39). But if the strata be softer and of more uniform texture, erosion yields rounded hills, often very steep, and sometimes passing into pinnacles, as in the Bad Lands of the west. Broad, open folds, as we have seen, give, normally, synclinal hills and anticlinal valleys (Fig. 22), when the erosion is well advanced. But in more strongly, closely folded rocks the ridges and valleys are determined chiefly by the outcrops of harder and softer strata, as shown in Fig. 40, the symmetry of the reliefs depending upon the dip of the strata. This principle of unequal hardness or durability also determines most of the topographic features in regions of metamorphic and crystalline rocks, in which the stratification is obscure or wanting.

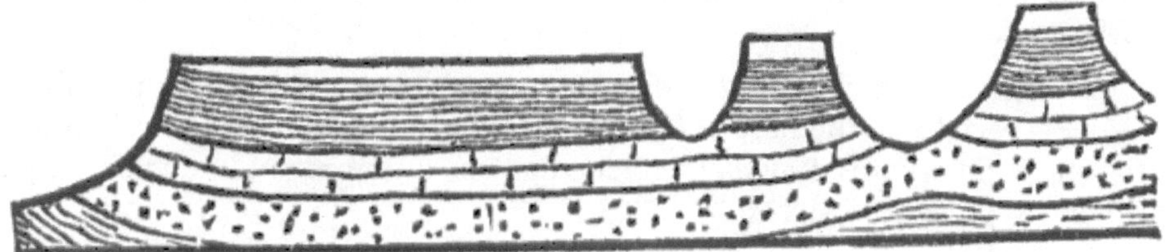

Fig. 39.—Horizontal strata and table-mountains.

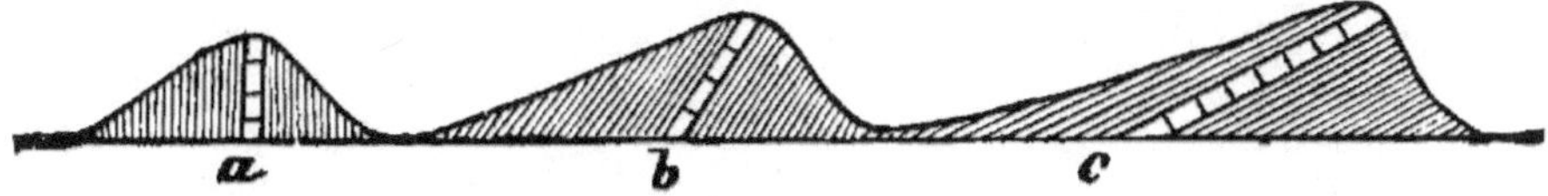

Fig. 40.—Ridges due to the outcrops of hard strata.

The boldness of the topography, and the relation of depth to width in valleys, depends largely upon the altitude above the sea; but partly, also, upon the distribution of the rainfall, the drainage channels or valleys being narrowest and most sharply defined in arid regions traversed by rivers deriving their waters from distant mountains. That these are the conditions most favorable for the formation of cañons is proved by the fact that they are fully realized in the great plateau country traversed by the Colorado and its tributaries, a district which leads the world in the magnitude and grandeur of its cañons. But deep gorges and cañons will be formed wherever a considerable altitude, by increasing the erosive power of the streams, enables them to deepen their channels much more rapidly than the general face of the country is lowered by rain and frost. This is the secret of such cañons as the Yosemite Valley, and the gorge of the Columbia River, and probably of the fiords which fret the north-west coasts of this continent and Europe. For a full description and illustration of the topographic types developed by the action of water and ice upon the surface of the land, and of the various characteristic forms of marine erosion, teachers are referred to the larger works named in the introduction, especially Le Conte's Elements of Geology, and to the better works on physical geography. We will, in closing this section, merely glance at some of the minor erosion-forms, which are not properly topographic, but may be often illustrated by class-room and museum specimens. Mere weathering, the action of rain and frost, develops very characteristic surfaces upon different classes of rocks, delicately and accurately expressing in relief those slight differences in texture, hardness, and solubility, which must exist even in the most homogeneous rocks. Every one recognizes on sight the hard, smooth surfaces of water-worn rocks. They are exemplified in beach and river pebbles, in sea-worn cliffs, and where rivers flow over the solid ledges. The pot-hole (page 17) is a well-marked and specially interesting rock-form, due to current or river erosion.

Ice has also left highly characteristic traces upon the rocks in all latitudes covered by the great ice-sheet. These consist chiefly of polished, grooved, and scratched or striated surfaces, the grooves and scratches showing the direction in which the ice moved.

The organic agencies, as already noted, accomplish very little in the way of erosion, especially in the hard rocks, but the rock-borings made by certain mollusks and echinoderms may be mentioned as one unimportant but characteristic form due to organic erosion.

APPENDIX

The following collections are especially prepared and arranged for use with this text:

Weathering

  1 Diabase

*2 Diabase weathered

*3 Diabase disintegrated

°4 Felsite: Angular fragment

°5 Felsite: Water rounded pebble

Formation of Coals

*6 Peat

°7 Lignite

  8 Bituminous

*9 Cannel coal

°10 Anthracite

°11 Native coke

Rock-forming Minerals

*12 Graphite

°13 Halite

*14 Limonite

*15 Hematite

*16 Magnetite

 17 Magnetite Lodestone

*18 Quartz: Glassy

 19 Quartz: Flint

 20 Quartz: Chert

 21 Opalized wood

*22 Gypsum

*23 Calcite

°24 Dolomite

25 Siderite
*26 Hornblende
°27 Pyroxene
*28 Muscovite
29 Biotite
*30 Orthoclase
°31 Albite
*32 Labradorite
*33 Kaolinite
34 Talc
*35 Serpentine
°36 Chlorite
37 Glauconite (Green Sand)
38 Chrysolite
39 Garnet
40 Pyrite

Sedimentary and Metamorphic Rocks
*41 Conglomerate: Breccia
*42 Conglomerate: Pudding-stone
*43 Sand: Quartz
°44 Sand: Magnetite
*45 Sandstone: Ferruginous
46 Sandstone: Calcareous
47 Sandstone: Arkose
*48 Quartzite
49 Clay: Boulder
°50 Clay: Fire
*51 Shale
*52 Shale Carbonaceous
53 Slate: Roofing
54 Slate: Flagstone
55 Porcelainite
56 Tripolite
°57 Siliceous Tufa
58 Novaculite

°59   Asphaltum

°60   Oil Sand

*61   Limestone: Fossiliferous

*62   Limestone: Coquina

*63   Limestone: Chalk

 64   Limestone: Crystalline

°65   Limestone: Compact

 66   Limestone: Hydraulic

 67   Calcareous Tufa

 68   Dolomite

 69   Rock Salt

°70   Phosphate Nodule

*71   Gneiss: Granitoid

*72   Gneiss: Micaceous

 73   Gneiss: Hornblendic

°74   Norite: Hypersthenite

*75   Schist: Mica

 76   Schist: Hornblende

 77   Schist: Talc

 78   Schist: Chlorite

 79   Amphibolite

 80   Soapstone

 81   Verd Antique (Serpentine)

Igneous Rocks

*82   Granite: Binary

 83   Granite: Muscovite

*84   Granite: Biotite

 85   Granite: Hornblendic

 86   Granite: Red

*87   Syenite

 88   Syenite Elæolite

*89   Diorite

*90   Diabase: Trap

*91   Rhyolite

 92   Trachyte

*93 Obsidian

94 Pumice

°95 Petrosilex

*96 Andesite

*97 Basalt

98 Basalt vesicular Lava

*99 Melaphyr: Amygdaloidal

°100 Volcanic Tuff

Collection No. F1. Entire list of 100 museum size specimens (3¼ × 4¼), numbered, labelled and mounted on blocks or in improved trays, for museum display and laboratory work $40.00

(The same, labelled but unmounted, $30.00)

Collection No. F2. Same as above, but small museum size, mounted in improved trays (2½ × 3½) $25.00

Collection No. F3. Same as F2, but hand size specimens (2 × 2) 12.50

Collection No. F4. 80 specimens, omitting those marked (°), in individual trays (2½ × 1¾) and two cloth-board cases, numbered to correspond with accompanying printed list (no labels) 5.00

Collection No. F5. 40 specimens marked (*), mounted as collection F4 2.50

Collection No. F6. 100 pupils' fragments (1 × 1), numbered, in paper bags. (Single collection $1.25.) In lots of 5 or more, each 1.00

Collection No. F7. 80 pupils' fragments (like F6). (Single $1.00.) In lots of 5 or more, each .75

Collection No. F8. 40 pupils' fragments (like F6). (Single 50c.) In lots of 5 or more, each .40

Collection No. F9. 25 museum size specimens, illustrating structure, faults, stratification, etc. Mounted and labelled 10.00

Lector House believes that a society develops through a two-fold approach of continuous learning and adaptation, which is derived from the study of classic literary works spread across the historic timeline of literature records. Therefore, we aim at reviving, repairing and redeveloping all those inaccessible or damaged but historically as well as culturally important literature across subjects so that the future generations may have an opportunity to study and learn from past works to embark upon a journey of creating a better future.

This book is a result of an effort made by Lector House towards making a contribution to the preservation and repair of original ancient works which might hold historical significance to the approach of continuous learning across subjects.

HAPPY READING & LEARNING!

LECTOR HOUSE LLP
E-MAIL: lectorpublishing@gmail.com